Resilience, Community Action

and Societal

Transformation

Edited by
Thomas Henfrey
Gesa Maschkowski
and Gil Penha-Lopes

Permanent Publications

Published by
Permanent Publications
Hyden House Ltd
The Sustainability Centre, East Meon, Hampshire GU32 1HR, United Kingdom
Tel: +44 (0)1730 823 311
Fax: +44 (0)1730 823 322
Email: enquiries@permaculture.co.uk
Web: www.permanentpublications.co.uk

Distributed in the USA by
Chelsea Green Publishing Company, PO Box 428, White River Junction, VT 05001
www.chelseagreen.com

Designed and Typeset by vivoeusebi

Printed in the UK by CPI Antony Rowe, Chippenham, Wiltshire

All paper from FSC certified mixed sources
The Forest Stewardship Council (FSC) is a non-profit international organisation
established to promote the responsible management of the world's forests.
Products carrying the FSC label are independently certified to assure consumers
that they come from forests that are managed to meet the social, economic and
ecological needs of present and future generations.

British Library Cataloguing-in-Publication Data
A catalogue record for this book is available from the British Library

ISBN 978 1 85623 297 5

Editorial input from Gil Pessanha Penha-Lopes and printing were financially supported by
Fundação para a Ciência e Tecnologia (FCT Unit funding, ref IUD/BIA/00329/2013; and
IF/00940/2015).

Creation of this book has been partially supported by funding from the
European Union's Seventh Framework Programme for research, technological
development and demonstration under grant agreement 308337 (Project BASE).
Its contents are the sole responsibility of contributors and can in no way be taken
to reflect the views of the European Union. http://base-adaptation.eu/

Table of Contents

List of Contributors

LORENZO CHELLERI

Chair, Urban Resilience Research Network;
Lecturer, International University of Catalonia;
Research Associate, Gran Sasso Science
Institute, Italy

MAY EAST

Chief Executive
Gaia Education

NARESH GIANGRANDE

Training Coordinator and
Research Coordinator
Transition Network, UK

MAJA GÖPEL

Head of Berlin Office
Wuppertal Institute, Germany

JANINA GRABS

Institute for Political Science
University of Münster, Germany

THOMAS HENFREY

Research Fellow
Centre for Ecology, Evolution and
Environmental Change, Lisbon University,
Portugal; Schumacher Institute for
Sustainable Systems, UK

JUSTIN KENRICK

Africa Regional Coordinator and
Policy Advisor
Forest Peoples Programme

GLEN DAVID KUECKER

Professor of History
DePauw University, USA

NINA LANGEN

Professor, Institute of Vocational Education
and Work Studies
Division of education for sustainable food
consumption and food science
Technische Universität Berlin, Germany

MELISSA LEACH
Director, Institute of Development Studies,
University of Sussex
Founding Director, STEPS Centre, UK

CHERYL LYON
Transition Town Peterborough, Canada

CHRISTOPHER LYON
Centre for Environmental Change and
Human Resilience, University of Dundee, UK

GESA MASCHKOWSKI
Institute for Food and Resource Economics,
University of Bonn, Germany

EAMON O'HARA
Executive Director
ECOLISE: the European network for
community-led initiatives on climate
change and sustainability

JUAN DEL RÍO
Transition Trainer
Coordinator Red de Transición,
Transition Spain

HELEN ROSS
Professor of Rural Development
The University of Queensland, Australia

NIKO SCHÄPKE
Institute for Ethics and Transdisciplinary
Sustainability Research,
Leuphana University of Lüneburg, Germany

BRIAN WALKER
Honorary Research Fellow
CSIRO, Australia

INTRODUCTION

1

1.0. Introduction: A Community-led and Knowledge-led Transition

EAMON O'HARA AND MAY EAST

As discussions continue on global and national responses to issues around climate change and sustainability, local communities around the world are busy taking action, leading the way towards what could be a revolutionary, community-led societal transformation. Weary of waiting for top-down solutions, and determined to be active participants in shaping the future, citizens are taking the initiative and launching transformative processes in their own neighbourhoods and villages.

A study carried out by the European Association for Information on Local development (AEIDL) in 2013 identified in excess of 2,000 community-led initiatives on climate change, resilience and sustainability operating across 13 EU Member States.[1] This number — probably a massive underestimate — is growing rapidly, spearheading a movement that has the potential to transform society, building local resilience and establishing new norms and behaviours that support the shift to a post-carbon society.

In many cases, these local initiatives are aligned with wider movements, in particular the Transition Towns network,[2] the Ecovillage movement[3] and the Permaculture movement.[4] Mostly, such initiatives adopt a systemic/holistic approach, operating across many different sectors and themes.

There are also many community-led initiatives that have a greater sectoral focus – for example on community-supported agriculture, community energy or community mobility - leading to the flowering of a diversity of practical projects. In all cases, however, the focus is on

1   http://www.aeidl.eu/images/stories/pdf/transition-final.pdf
2   http://www.transitionnetwork.org
3   See http://gen.ecovillage.org and http://gen-europe.org
4   e.g. see http://permaculturenews.org and http://www.permaculture.org.uk

experimentation, innovation, and on finding practical and viable ways of strengthening local resilience and sustainability.

In Europe, a number of these movements and initiatives have recently come together to establish a platform to support closer cooperation. This European meta-network for community-led action on climate change and sustainable development, known as ECOLISE, aims to raise the profile of community-led action, among policy-makers and the general public, with a view to promoting a wider mainstreaming of such action across Europe.[5]

Three key concepts inform the growing number of citizen-led initiatives: localisation, resilience and inventiveness. Localisation invites us to rethink methods of production and patterns of consumption, arguing that meeting more needs locally while plugging the leaks in our local economies will be one of the critical strategies for improving the resilience of our communities. On the task of strengthening resilience within and across systems, we observe the adaptability and transformative potential of community-led initiatives as we notice the richness and complexity of the socio-economic webs that unfold, suggesting that innovation happens in the intersections between the 'movements'. Working in isolated enclaves may work if the world around you is not changing. ECOLISE's intention of bridging 'clusters' of practical post-carbon experimentation is already creating cross-fertilisation; inventiveness will certainly follow.

The link between researchers and practitioners is key in this respect. An established body of knowledge now exists, providing a wealth of good practices and processes that can be transferred to mainstream society, helping to fast-track a community-led societal transformation. However, to be truly successful, this mainstreaming requires a thorough understanding of the underlying processes at play within active communities and their interaction with social, economic and political factors at higher levels.

To develop the right policies and to promote the most effective approaches, it is imperative that we understand what works best in what context and why. Close working relationships between researchers and practitioners are essential to achieving this. Through this kind of cooperation, a more scientific approach is possible, facilitating ongoing two-way processes of learning and improvement and creating a framework through which experience can be capitalised and shared.

In this context, the mission of ECOLISE has two important components. Firstly, to map and investigate what's currently happening in terms of community-led action on climate change and sustainability in Europe, what works and what doesn't, and to share this knowledge throughout the network. Secondly, to raise the profile of community-based action on climate change and sustainability in Europe and promote wider sharing and uptake of the best ideas and experiences emerging from the network.

5   http://www.ecolise.eu

We all recognise the need for urgent action on climate change and sustainability. Time is of the essence in terms of avoiding the worst effects of climate change, halting the loss of biodiversity and protecting finite resources. This is why it is so important that we acknowledge and support community initiatives, and encourage more local communities to take action. However, to facilitate this we must also push for a more supportive policy environment, and encourage policymakers, especially at the European and national levels, to bring forward policies and programmes that support community-based action. We already know there are politicians and bureaucrats who support such efforts but we must work together to present a more coherent and science-based proposition in order to bring about real change.

1.1. Editorial and Summary

TOM HENFREY

This book explores the interface, or edge, between resilience research and various grassroots resilience-building initiatives arising within the last few decades in reaction to sustainability and related challenges and the perceived inadequacy of centralised responses on the part of government and business. It gathers together relevant materials from the Resilience 2014 conference, organised by the Resilience Alliance, which took place in Montpellier, France from May 5th–9th 2014. Many are derived from a parallel session organised by activists in the Transition movement; it also compiles scattered contributions from elsewhere. They provide insights into what may well be a highly significant phase in the history of community action on resilience, which is potentially on the cusp of its own transition: from a marginal and largely experimental endeavour at the fringes of dominant political and economic systems and restricted by what they permit, to a significant force for global change that reshapes these systems in ways that reflect the long-term interests of the world as a whole, not misguided greed, insecurity and self-interest.

1.1.1. Context

Transition is one of many such grassroots responses, and originated in the market town of Totnes in South West England in 2006. It builds on ideas developed in a permaculture course taught by Transition founder Rob Hopkins at Kinsale Further Education College in Ireland the previous year, whose final project applied permaculture to designing a community-led response to peak oil. In the terms of Resilience Theory's adaptive cycle model (explanations of technical terms in this and the next few paragraphs can be found in Chapter 3.2.2), Transition can be viewed as emerging from a reorganisation phase in the permaculture community resulting from the change in broader context as peak oil went from an abstract possibility to directly experienced reality. Having added climate change to its core concerns, and in the context of high levels of public concern and political rhetoric about climate change, Transition entered what Resilience Theory terms an r phase, rapidly growing to become a worldwide movement that at the time of writing in summer 2014 includes over 1000 registered local initiatives in at least 40 countries worldwide.[6]

6   See _http://www.transitionnetwork.org/initiatives/_

Recent years have shown indications that Transition is entering a K phase – one of relative stability, marked by more or less established norms and patterns of activity[7] – or a release phase in which resources and energy within the movement dissipate and/or decline. Creation of new initiatives has plateaued in the UK, where the movement is most mature (although numbers continue to increase elsewhere, and overall). Many UK-based initiatives report stagnation or reduced levels of activity, and some have ceased to operate in any effective sense. Transition Network, the Totnes-based charity that acts as a coordinating and support body for the movement, has entered a significant phase of reorganisation of its entire structure and operations, responding to the dramatic changes in context since 2006 and incorporating collective learning within the movement over that time.

Reorganisation phases are critical for any system, and in this case present a risk that the Transition movement either vanishes without trace or persists inconsequentially in the background of a world system that does its best to maintain business as usual. Many regarded the financial crisis of 2008-9 as both compelling and providing the opportunity for fundamental rethinking and reorganisation of the global financial system, and indeed of our economy and way of life. High level political action has instead been dominated by the 'Remember' effect, focussing efforts on restoring and maintaining economic growth rather than heeding widespread calls for deeper transformation.[8] The remember effect manifests in people's lives

7   A summary of these from 2011 can be found at *http://www.transitionnetwork.org/ingredients*

8   Jackson, T., 2009. *Prosperity Without Growth*. London: Earthscan. Heinberg, R., 2011. *The End of Growth*. Forest Row: Clairview.

Figure 1.2.1. – *Brussels, May 2014 - Ecolise Registration.*

when Transition initiatives disappear without any significant and lasting effects on their local economy, and people within them return to lives, lifestyles and livelihoods only superficially different, if at all, from how they were before. The same would be true if Transition, as a movement, were to compromise its basic commitment to helping bring about a sustainable and fair society, and instead join the ranks of reformist civil society movements who dull some of the rough edges of predatory capitalism without challenging its central social and cultural role. Fortunately, Transition's reorganisation phase can take advantage of significant 'Revolt' effects where grassroots action has changed policy environments – and discourses within them and related areas – in significant ways. Transition and similar grassroots movements have to certain extent created their own "context for renewal"[9] more conducive to their continued development. This is an example of the permaculture principle 'everything gardens': in other words, resilient systems change their own environments in ways that benefit both them and other interdependent systems. These changes are manifest in marked tensions between rhetorical commitment to action on climate change and other sustainability issues, and entrenchment of macro-economic policies that reinforce and exacerbate their causes. Many discussions in parallel

Initiatives seek to transcend the contradictions arising from the need to work within existing systems at the same time as they render them obsolete.

and plenary sessions at Resilience 2014 reflected this, drawing increasing attention to the issues of power that determine how – and in whose interests - resilience is defined and operationalised. For community-based practitioners working towards fairer and more sustainable societies, resilience means something very different from what it does to governments and businesses seeking to entrench their own privileged position within the status quo.

Attention to these different framings of key guiding concepts is one example of how grassroots movements are becoming increasingly conscious of the importance of cross-scale interactions - in other words, working within panarchies rather than delimited local contexts. Debate within Transition about its relationship with power and power structures increased in intensity and sophistication within the past year or so, and reflects some of the most important learning the movement has achieved in its short history. Closely linked are new

9   Folke, C., J. Colding & F. Berkes, 2003. Synthesis: building resilience and adaptive capacity in social-ecological systems. Pp. 352-387 in Berkes, F., J. Colding & C. Folke (eds.). *Navigating Social-Ecological Systems*. Cambridge University Press.

relationships with research and researchers that emphasise the need to subvert traditional power relationships and develop more inclusive and equitable working methods. Initiatives like the Transition Research Network,[10] Research in Community,[11] and the UK Permaculture Association's research programme[12] promote and enact overtly politicised forms of transdisciplinarity that draw on the social and cultural knowledge the permaculture, ecovillage and Transition movements have developed over the years.[13, 14, 15] By making explicit the need for radical and transformative action, and the importance of mobilising resources in different ways at different levels, such initiatives seek to transcend the contradictions arising from the need to work within existing systems at the same time as they render them obsolete.

> *The experience of emergence is the magic at the heart of Transition and permaculture, tools for genuine resilience of the kind that can only arise in autopoieitic fashion.*

An important new manifestation of this is the establishment of ECOLISE, the European Network of Community-Led Initiatives on Sustainability and Climate Change. Deriving in part from international networks within and between the Transition, permaculture and ecovillage movements, in part from initiatives for collaborative research, and in part from the desire for advocacy at various levels of government, ECOLISE grew out of a process initiated by AEIDL, the European Association for Information on Local Development, who in 2013 produced an initial summary report[16] and assembled interested parties for a planning meeting in Brussels. On Friday May 9th 2014 — the day after the Resilience 2014 Conference finished in Montpelier, ECOLISE was formally constituted at a legal ceremony in Brussels by 25 founding organisations representing Transition initiatives, ecovillages, permaculture projects, and other grassroots initiatives for promoting resilience through community-based action, along with municipal initiatives and specialised organisations dedicated to supporting such action through work in areas such as research and education.

10   http://www.transitionresearchnetwork.org

11   http://www.researchincommunity.net/

12   http://www.permaculture.org.uk/research

13   Henfrey, T., 2014. Edge, Empowerment and Sustainability: Para-Academic Practice as Applied Permaculture Design. In The *Para-Academic Handbook: A Toolkit for making-learning-creating-acting*. London: HammerOn Press.

14   Sears, E., C. Warburton-Brown, T. Remiarz & R. S. Ferguson, 2013. A social learning organisation evolves a research capability in order to study itself. Poster presentation at the Tyndall Centre Radical Emissions Reduction Conference, London, UK, 10th –11th December 2013.

15   Andreas, M. & F. Wagner (eds.), 2012. Realizing Utopia: Ecovillages and Academic Approaches. RCC Perspectives 2012/8. Munich, Rachel Carson Center for Environment and Society (RCC).

16   http://www.aeidl.eu/images/stories/pdf/transition-final.pdf

Figure 1.2.2. – *Workshop Organisers and Speakers.*

1.1.2. Content

The core of the present book comes from a parallel session at the Resilience 2014 conference on Tuesday May 6th 2014, which in part sought to celebrate the foundation of ECOLISE and strengthen its links with the international resilience research community. Most of the session's organisers had central roles in the establishment of ECOLISE over the previous two years, particularly in developing an interface between research and community action that is one of the core pillars of ECOLISE. Introducing and facilitating the session was Tom Henfrey, a coordinator of the Transition Research Network and Researcher at the Schumacher Institute for Sustainable Systems in Bristol, UK. Three short presentations then set the scene; each a shorter version of a chapter in the present volume. Juan del Rio of the Spanish national Transition hub reported on a research collaboration with Lorenzo Chelleri of the Barcelona Autonomous University that documented the early stages of Transition in Spain (Chapter 2.2). Gesa Maschkowski, a PhD researcher at Bonn University and member of the German national hub, reported on research on the social conditions for effective community action, how Transition activists cultivate personal resilience in order to create conducive environments for behaviour change on the part of others (Chapter 3.1). Glen Kuecker, history professor at de Pauw University in Indiana and active supporter of indigenous struggles in Latin America presented a call for solidarity between Global North activists for social change and those fighting against injustice in the Global South (Chapter 4.3). The main part of the session employed Open Space, a social technology for resilience building that activates the collective capacity of a community to identify and interrogate key issues arising in any area of interest. Within the general topic of resilience and community action, all participants at the session were invited to propose questions for general discussion. The group agreed on four final questions and formed discussion groups around each of these, everyone present joining the group whose question most interested them. After the discussion, each group fed key points back to the group as a whole. The findings, a distillation of the collective intelligence of the group as a whole, form a chapter of their own (Chapter 4.1).

Other chapters come from relevant sessions elsewhere in the conference: Cheryl Lyon's account of Transition in Peterborough, Canada (Chapter 2.1) and Maja Göpel's paper on paradigm shifts and societal transformation (Chapter 3.3). To add theoretical context, we included a revised version working paper on Transition and Resilience developed by the Transition Research Network in 2012 to support work into monitoring and evaluation methods for Transition groups and not previously made publically available (Chapter 3.2). I also added to Section 2 an account of relevant activity in Bristol, UK, my home city at the time of the Montpellier conference, where grassroots and top-down initiatives on both resilience and sustainability are intersecting in interesting ways (Chapter 2.3). In addition, while at the conference we requested short contributions from several leading names in the resilience field whose contributions to plenary or other major sessions particularly resonated with the theme of the book, and which start the main sections in this collection: Melissa Leach (Chapter 2.0), Helen Ross (Chapter 3.0) and Brian Walker (Chapter 4.0). Finally, we added an essay that, complementing Glen Kuecker's, examines Transition in global context as part of a wider global commons movement (Chapter 4.2), originally written alongside the Transnational Institute's important collection on climate security, 'The Secure and the Dispossessed', here (in the paper version of this book) appearing in printed form for the first time.

1.1.3. The Magic of Emergence

Compiling and editing this collection has been an exciting and rewarding undertaking. To my surprise, it became a parallel to the incomparable experience of resilience as an emergent property of a system, arising of its own accord but that can be encouraged by careful nurturing. As it grew from a simple session write-up, new ideas and contributions appeared, intersections and synergies became evident among the different pieces and perspectives they represent. The whole took on a life of its own, and with no effort at centralised control or organisation, took on a coherence among its constituent chapters that makes them far more than the sum of its parts. My role gave me an experience of emergence that is hard to express in words, but which will be familiar to anyone who has experienced being at the heart of a spontaneous unfolding of this type. This is the magic at the heart of Transition and permaculture, tools for genuine resilience of the kind that can only arise in autopoieitic fashion.

VIEWS FROM
THE GROUND

2

2.0. Citizen-led pathways to sustainability
MELISSA LEACH

Meeting today's sustainability challenges requires structural changes that move economies and societies away from business as usual, and onto greener and fairer paths. The changes needed are not just technical and economic, but fundamentally political. Politics and power are important to how pathways are defined and shaped, towards which versions of 'sustainability' and 'green'; which pathways win out and why, and who benefits from them – how far they serve issues of justice and fairness.

Figure 2.0.1 – Melissa Leach. Credit: Gesa Maschkowski.

The politics of transformations to sustainability must embrace more fully community resilience and empowerment, and citizen knowledge, action, capacities and mobilisations. These can contribute to transformations from below, in ways that vitally complement technology-led, market-led and state-led approaches.

Citizen-led and grassroots innovation show that valuable solutions do not just originate in the hi-tech laboratories of firms and technology start-ups in the global North or the emerging economies of China, India or Brazil. Instead, in cities and rural settings, networks of individuals, development workers, community groups and neighbours have been generating technological and social innovations in sectors as diverse as water and sanitation, housing and habitats, food and agriculture, energy, mobility, manufacturing, health, education, communications, and many others. Examples include the thousands documented by the *Honey Bee Network* in India, now supported by government through the *National Innovation Foundation*.

There are also citizen-led alternative economies and mobilisations around them, including transition towns, alternative agri-food and food sovereignty movements. Civil society groups have also proposed alternative ways of 'living well', such as plans for *Buen Vivir*, now endorsed by government ministries in Ecuador, that combine environmental justice, common goods, agro-ecology and food sovereignty.

Such citizen innovation and mobilisation processes are frequently motivated by a mesh of socio-cultural and livelihood concerns - and understandings of ecology and sustainability – that diverge from the narrow notions of 'green', 'environment' and 'economic benefit' encompassed in more technocentric sustainability and green economy discourses.

Far from being just confined to the local, citizen and community perspectives and action illustrate and offer broader contributions to the politics of green transformations.

First, they embrace a **politics of knowledge**: they emphasise diversifying and democratising knowledge for transformations, beyond official expertise and formal science, to include experiential, informal and indigenous expertise. Attention to citizen knowledge can help to 'open up' discussions, allowing for discursive reframing, and deliberation and dialogue as part of a process of knowledge production for and within transformations.

> *Civic action to disrupt, discontinue and challenge incumbent power, and offer alternatives, has always been a central part of historical transformations.*

Second, through **networking**, green citizen and community action often moves well beyond the local. Thus we have seen the potential of place-based struggles to resonate and 'globalise' through transnational advocacy networks in examples from movements like agro-ecology and food sovereignty, or those around carbon justice. The challenge is to retain a balance of autonomy and groundedness, and the vital traction on people's hopes, fears and energies, with scale-up and wider institutionalisation.

Third, citizen movements **challenge and shape powerful structures,** often with other actors, becoming part of an important politics of alliance building. For example, activists have sought to mobilise the power of finance capital given its heightened power to drive decarbonisation. Such civil society challenges often emphasise justice as well as greenness; or at least versions of green that are also socially just. Thus mobilisations for alternative pathways in which rights to food, water or energy often have a central role, are combined with resistance to existing forms of extractivism and business-as-usual development.

We therefore need to include citizen perspectives and actions more centrally across the ways we articulate and pursue pathways to sustainability and green transformations.

Figure 2.0.2 – *French Organic Farmers Working on Land Funded by Tierre de Liens. Credit: Gesa Maschkowski.*

Civic action to disrupt, discontinue and challenge incumbent power, and offer alternatives, has always been a central part of historical transformations – whether the ending of slavery, or feminism - and will continue to be part of future ones. As the past also tells us, transformations will be messy – although with hindsight they may look like planned linear change or be imagined as such, the reality is always contested, overlapping strategies and alliances. And they will be context specific – given the diversity of accumulation strategies being pursued by states and corporations in different parts of the world and the ways in which they enrol and collide with so many other social actors, we can expect a diversity of pathways, and should be wary of 'blueprints', 'models' and 'transfers' from 'success' stories. These politics will continue to play out on a terrain of competing discourses, institutions and material interests in diverse contexts. The challenge for all of us is to engage on that terrain in defining and realising pathways that are both green and just.

2.1. Transition in Peterborough, Canada
CHERYL LYON AND CHRISTOPHER LYON

This chapter reports on my experiences of the challenges to resilience-building in a mid-sized Canadian city.

2.1.1. Local Context
The City of Peterborough, Canada was first called *Nogojiwanong* ("where the rapids end"), and lies in the Mississauga Territory of the original Anishnaabeg people.

Our population is 76,000 people, 130,000 including surrounding rural area and villages. Our geographic region was shaped by retreating glaciers into rolling drumlins and eskers generously endowed with lakes and rivers. Some logging and trapping still occur in the northern part of our area, farming in the southern. We are a holiday destination for many. There are three indigenous communities in our vicinity. Our university, Trent, is well-known in Canada for Indigenous Studies. Our economy is a mix of a declining manufacturing sector (with a high percentage of that workforce in 'Creative Class' occupations) and a large number employed in service industries and education.[17] Demographically, we have the highest population of seniors in Canada. We also face the most serious housing unaffordability in Canada and, at times, the highest unemployment rate.[18]

2.1.2. Personal Bio and Transition Town Story
Myself, I have a small consulting business in the arts of group facilitation. I have also worked in municipal government in Emergency Planning and Housing administration, in not-for-profit direct service social welfare agencies, and at both the Federal and Provincial levels of government.

I volunteer with **Transition Town Peterborough** (TTP). The Transition Town movement is voluntary, not-for-profit groups of local citizens with a passion for introducing resilience

---

17   Who Works Where: Occupational and Industrial Distribution in Ontario. Martin Prosperity Institute, Rotman School of Management, University of Toronto, Dec. 2013.

18   Canada Mortgage and Housing Corporation stats 2012.

practices in their communities to adapt to the interconnected effects of climate change, the end of cheap fossil fuel energy, and economic contraction. Peterborough was Canada's first Transition Town.

2.1.3. Transition Town Peterborough's Challenges

At Resilience 2013, Suzanne Moser, in her summary mirroring poem 'Maybe...' included a conference participant's phrase *"here be dragons"* to describe the unknown, frightening territory off the edge of familiar maps in the Middle Ages. I've borrowed that image to an extent as a way of naming the **challenges** we face inviting people into the new territory that Transition is exploring: the profound changes of worldview, ideas of comfortable practices and habitual expectations now necessary due to the impacts of climate change.

In one sense, our timing couldn't be worse!

> *"just when we needed to gather, our public sphere was disintegrating; just when we needed to consume less, consumerism took over virtually every aspect of our lives; just when we needed to slow down and notice, we sped up; and just when we needed longer time horizons, we were able to see only the immediate present."*
> (Naomi Klein, rabble.ca April 22, 2014)

Challenge #1: How do we get round the socio-cultural-economic privileging of unsustainable lifestyles: the addiction to the privileges and comfort of an affluence based on fossil fuels?
This powerful addiction has everyday social and cultural reinforcements that reach us in the pleasure and security/safety centres of our brains. The prospect of giving up even some of these privileges is too scary or unbelievable, and is denied. A local example of addiction reinforcement: one of Peterborough's largest employers, employing some 2000 Peterburians, is a General Motors automobile manufacturing plant, a 45-minute commute away.

> *We do not share the deep-seated, historically-reinforced belief and trust in government power and efficacy to act for the good of all and to protect us from the excesses of the marketplace.*

TTP's response to this challenge is education and awareness-raising (e.g. through films and discussions), and our quarterly free magazine called the *Greenzine*.

Challenge #2: How do we change deference toward and trust in government and big business into trust in personal responsibility and agency as sources of action?
TTP emphasizes grassroots, community-building action for adaptation. We encourage

Figure 2.1.1 – *Transition Town Peterborough. Credit: POF.*

individual and small group initiatives. While supporting a vital role for governments, we do not share the deep-seated, historically-reinforced belief and trust in government power and efficacy to act for the good of all and to protect us from the excesses of the marketplace. Successive Canadian senior governments have shifted towards more neo-conservative values in which politics serves and protects the growth-based, fossil fuel extraction economy. Government-business partnerships and technological innovation are often looked to for economic solutions.

TTP's response to this challenge is to work at it (as we do all the challenges) *bit by bit* across several initiatives that are examples of alternatives: by just doing it. A key future initiative of this type is to start neighbourhood-based **Resilience Circles**, also called **Transition Streets**, starting small in building resilience through face-to-face interaction among neighbours. This will require planning and resources we don't have right now.

Challenge #3: the Conventional Focus of Municipal Responsibilities: *potholes versus preparation.*
How does TTP bring municipal government into true adaptive planning when its main focus is still new roads, keeping taxes low, setting rules for bicycles and attracting big new employers?

The City Emergency Plan is designed, "to ensure the co-ordination of municipal, private and volunteer services in an emergency to bring the situation under control as quickly as possible." There is no accounting for prolonged emergencies of magnitudes not seen before, for social unrest, prolonged infrastructure damage, paralysis of senior government aid, long-term business stoppage, fuel shortages, interruption of food supply, or other situations that, although unprecedented, are increasingly likely.

Some things we've done in response are:

› to become the first Champion-level Partner with the City's official *Sustainability Plan* in order to bring our ideas forward
› to meet one-on-one with City Councillors for the same purpose and (similar to previously mentioned responses)...
› To offer education and conversation opportunities to learn about banking, money and economy (e.g. film series) and our free quarterly *Greenzine*.

We also search out and seize every opportunity to talk about what we do. Examples include a conference on Urbanism for International Development Students and an information booth at the local Spring Garden Show. We are also ramping up our efforts to use Social Media for information purposes, and host monthly "Meet Ups" open to all at a local pub.

Challenge #4: How to break through a local sense of immunity to world problems?
Peterborough's is a relatively benign climate. Other than a sudden flood in 2004 - the result of a rare, local weather anomaly - we have had no disastrous climate events with long-lasting, disruptive impact. Local government carries on economic development under a 'business-as-usual' model, investing tax reserves in big banks and not convening the vital public conversations on our future. The prevailing story is that Canada is immune to what's happening internationally. Many Canadians feel they weathered the last recession relatively well because government tightly regulates our Big Banks. Our consumer debt load is astronomical. On top of this, there is a local lack of knowledge and/or acceptance among the general populace and politicians of what is coming due to climate change.

Our response to this challenge is, again:

› to offer education and conversation opportunities to learn about banking, money and economy (e.g. film series) and our free quarterly *Greenzine*.

We also

› search out and seize every opportunity to talk about what we do e.g. a conference of IDS students on Urbanism and the local Spring Garden Show
› are ramping up our efforts to use Social Media
› hold monthly "Meet Ups" open to all at a local pub.

Challenge #5: How do we overcome a focus on single issue interests and pre-determined outcomes?

Potential players in adaptive responses are often preoccupied with their particular turf and planned outcomes in an unpredictable future. Businesses are focused on

> *TTP tries to "connect the dots" in everything we do among various sectors and players in the nested, dynamic process that is adaptation and resilience-building.*

the bottom line and outright day-to-day survival. Social support agencies have single-issue mandates and silos of funding. The transition challenge is how to connect them *together* in concerted effort to survive and adapt when *everything* is now driven by climate change in a world where outcomes are less predictable and new possibilities need to be imagined.

In response, TTP tries to "connect the dots" in everything we do among various sectors and players (business, human services, public and private) in the nested, dynamic process that is adaptation and resilience-building.

Challenge #6: Against this backdrop, you may appreciate TTP's challenge of conveying the ideas of transitioning to a resilient community, particularly in our main initiative: Economic Localization.

In general, Economic Localization gives preference to local resources, creates employment in locally-owned enterprises, creates local conditions and instruments for giving credit, and strengthens the non-monetized economy, seeking less dependence on external imports.

TTP's concept of *localization* is difficult for a populace who generally do not understand the meaning of money and the economy, or the role of the banking system. It is challenging to engage citizens in a grassroots movement when they may be inexperienced in terms of previous participation in change efforts or disenchanted by previous attempts via the political system.

TTP responds to the localization challenge in two main ways:

First, it *distinguishes itself from local environmentalism* by this: climate change action plans and energy descent efforts will not be enough unless we underpin them with a local *economic infrastructure* as a key adaptation to climate change. Plans for lower carbon footprints, more green spaces, increased housing density, green building forms, active transportation, food from with 100km, and so on, are all necessary, but if we cannot continue to pay for, produce and supply our life essentials (food, water, energy, culture and health), cannot continue to trade among ourselves in ways that support livelihoods – then we will not be truly sustainable or regenerating communities.

Second, it is trying *to build a local economic infrastructure* in 2 main ways:

i. Our highest profile economic localization effort is our **Local Currency** called the Kawar-tha Loon. Over 100 local business now accept it (including a Mitsubishi car dealership). The Loon is a tangible symbol of keeping local wealth circulating in the community.
ii. We are working toward a **Public Trust**, a fund, seeded by the Canadian dollar reserve built up by the exchange of national currency for local currency. Hopefully, this will leverage other, bigger funds (we're aiming for municipal tax revenues) for investment in energy and the environment.

These are two of what economist Milton Friedman calls the 'good ideas' we want to have well-developed and 'lying at hand'[19] for when the real trouble begins.

Challenge #7: How do we ensure that Economic Localization benefits socially and economi-cally poor and marginalized people? What will our vision mean for people mired in structural poverty created by climate degradation, government-supported globalization and our own governments' social policies and austerity budgets? Will Economic Localization help alleviate poverty?
One of our responses to this challenge is Community Capacity-Building by:

› A very successful **Transition Skills Forum** for *reskilling* ourselves in essential goods and services (like bread- and cheese-making, repairs, green building, gardening etc.) using local knowledge while minimizing costs through the use of citizen volunteers whom we invite to bring their skills forward for the community. We make these and all our events as acces-sible as possible by charging only $5 or Pay What You Can, and holding them in physically accessible places.
› We partner and network with other local bodies (e.g. a credit union as our first banking agent for the local currency).
› We address the 'spiritual/sacred/emotional' in coming to grips with climate change in our **Heart and Soul** events.
› We also invite the community to come together in celebration and showcasing of local abundance and entrepreneurs (especially in the food sector) in our annual Purple Onion Festival and 'Live Local/Buy Local' exposition for local entrepreneurs.

A second part of our response to this challenge is to demonstrate, over time, that a local currency and community/public trust will foster trade among ourselves and keep small enter-prises (the backbone of any community economy) in business - hiring people into jobs that may not pay six-figure salaries but do have dignity, and pay enough for decent living – and in this way contributing to a self-perpetuating cycle of renewable, sustainable commerce that supports itself. In this scenario, employment is expected to continue to be *gradually* created: not in the usual way of Big Company descending saviour-like on the community with 200 jobs,

19   *http://theunbrokenwindow.com/2011/04/29/the-problem-with-economics/* Retrieved April 30, 2014.

only for those 200 jobs disappear all at once when it moves to take advantage of cheaper labour opportunities.

2.1.4. Summary

Although I began by saying "our timing couldn't be worse", I believe our timing also couldn't be better:

There is a rebirth of movement into unselfish, collective actions; into the offering of gifts; into work from the heart; into circles and spaces to explore in uncertainty and freedom in distributed creativity.

We are learning to ask new questions instead of having all the answers.

In summary, we are transitioning into resilience through:

> › love of our community
> › working in small groups (though a certain scaling up is also necessary)
> › as much localization as possible in life essentials
> › building momentum and creating demand for sustainable policy and practice
> › inner resilience (of Heart and Soul)
> › nudging away from competitive individualism to supportive community
> › strategic alliances wherever we can
> › being passionate, not scary, in our urgency to adapt to climate change
> › believing in what we are doing and just doing it!
> › and lastly - *Illegitimi non carborundum!*

2.2. Transition in Spain: A First Assessment of Dimensions, Challenges and Opportunities for Transition Town Initiatives

LORENZO CHELLERI AND JUAN DEL RIO

2.2.1. Introduction

Transition Initiatives (TIs henceforth) are increasingly present in Europe and worldwide, arising in urban, peri-urban and rural contexts as community led initiatives which self-organize responses to the most relevant and challenging long term societal issues: climate change and oil dependency (among others). Spain has recently become part of the international Transition movement and seen the establishment of several different initiatives, probably as a response to the dramatic economic crisis and its social consequences.

This chapter is a descriptive synthesis of the results of the first national survey of Transition Initiatives in Spain, the research project 'Barriers and Opportunities for Building Resilience: A Critical Assessment of Transition Initiatives in Spain'.[20] It includes insights from the research project's closing workshop *'Barreras y oportunidades para las iniciativas sociales hacia la sostenibilidad – Creando puentes en tiempos de cambio'*,[21] (Barriers and opportunities for social action towards sustainability: building bridges in changing times), held in Barcelona on February 21st 2014. It aims to provide as clear as possible a view on how Transition is happening in Spain: how different groups have been arising, organizing activities and interacting with people, institutions and among themselves.

From 2008, TIs emerged in several places, including Barcelona[22], Zarzalejo[23] (Madrid), and Coín[24] (Málaga). Within a few years their numbers rose quickly and in 2011 a group of representatives decided to organize the first and successful Encuentro Anual de Transición (annual national meeting), which took place in Zarzalejo (Madrid) in April 2012.[25] There, they launched the Red de Transición España (REDTE),[26] a national hub with a website for connecting, coordinating and sharing knowledge among the many different initiatives, for establishing links between the self-organized Spanish initiatives and the international Transition network, and for offering resources and trainings to help initiatives emerge and go forward.

While this network was being created, the need for some preliminary assessment of the numbers, sizes, networks and activities of TIs became apparent. This led to development of the 'Barriers and Opportunities' research project already mentioned, as a bridge between science and practice based on collaboration among the Autonomous University of Barcelona, the Venice IUAV University and members of TIs, aiming to answer these questions. The initial assessment was launched during the second Encuentro Anual de Transición, held in Mijas[27] (Málaga), where volunteers began interviewing meeting participants and distributing questionnaires, and subsequently completed by Transition Spain. Over the following eight months the project investigated group dynamics. The results were presented in Barcelona during a workshop in which a deeper discussion on barriers and opportunities for these movements took place.

---

20   *http://www.transitionresearchnetwork.org/transition-in-spain.html*

21   *http://www.transicionsostenible.com/wp-content/uploads/2014/01/Flyer-Taller-Barreras-y-oportunidades-de-las-iniciativas-sociales-para-la-Sostenibilidad.pdf*

22   *http://barcelonaentransicio.wordpress.com/*

23   *http://zarzalejoentransicion.blogspot.com.es/*

24   *http://www.coinentransicion.com/*

25   See *http://www.transicionsostenible.com/en-los-bordes-del-i-encuentro-iberico-de-transicion.htm* and the video of the first meeting at *http://www.youtube.com/watch?v=29rhEIW0zQU*

26   *http://www.reddetransicion.org*

27   Ver crónicas del 2° encuentro en *http://www.transicionsostenible.com/movimiento-de-transicion-musica-y-alegria-para-la-sostenibilidad.htm*

Figure 2.2.1 – Poster for the first Spanish Transition Gathering in April 2012. Credit: LaRana Gráfica.

Similar community-led movements in degrowth, ecovillage and permaculture have over the same time been growing in Spain. Although this first assessment limited itself to dynamics of Transition Initiatives, it has also has served as a preliminary investigation of strengths and weaknesses, identities, synergies and barriers of and among a broader range of groups, which together constitute Spain's broader transition movement.

2.2.2. How Many, Where and Who are They?

Various factors make it difficult to say precisely how many initiatives there are in Spain. Many arose from informal groups of a few friends, others from within larger pre-existing initiatives already established as a local association which started a Transition Initiative as a sub-project. Some have no internet presence and few or no connections with the national network, making it difficult to know about their existence; for some other groups is difficult to know whether they remain active. Finally, many groups attending the national Transition meeting and organized around related goals of sustainability and social cohesion, for various reasons choose not to self-identify as Transition initiatives.

However, we are able to give an initial picture of how the Transition movement looks in Spain. As Figure 2.2.2 shows, 57 transition initiatives appeared between 2008 and 2014. The map omits a number of other community-led initiatives, similar to TIs, that have been identified.

Figure 2.2.2 *– Map of Transition Initiatives Arising in Spain from 2008 to 2014. Credit: Red de Transición España.*

Table 2.2.1 provides a complete list of the Spanish TIs that appeared from 2008 to 2014, along with their locations and web page (when available). Table 2.2.2 lists other Spanish community-led initiatives similar to Transition initiatives.

Table 2.2.1 *– Spanish Transition Initiatives founded from 2008-2014, their locations and web addresses.*

SPANISH TRANSITION INITIATIVES		
Name	**Location**	**Web page**
Alaior en Transició	Menorca	*www.facebook.com/pages/ Alaior-en-Transicio/392393444170884*
Albacete en Transición	Albacete	*www.albaceteentransicion.blogspot.com*
Alhama en Transición	Alhama de Murcia (Murcia)	*alhamaentransicion.wordpress.com*
Álora en Transición	Álora (Málaga)	*www.facebook.com/AloraEnTransicion/*
Alozaina en Transición	Alozaina (Málaga)	
Alto Tiétar en Transición	Comarca del Alto Tiétar (Ávila)	*n-1.cc/g/alto-tietar-en-transición*
Argelaguer en Transició	Argelaguer (Gerona)	*www.argelaguerentransicio.com*
Axarquía en Transición	Comarca de Axarquía-Vélez Málaga (Málaga)	*www.facebook.com/groups/261011497363317*
Barcelona en Transició	Barcelona	*barcelonaentransicio.wordpress.com*
Barrio Alcosa en Transición	Barrio de Alcosa (Sevilla)	

SPANISH TRANSITION INITIATIVES		
Name	**Location**	**Web page**
Bilbao en Transición	Bilbao	*www.facebook.com/BilbaoEnTransicion*
Butroi en Transición	Mungia (Vizcaya)	
Carcaboso en Transición	Carcaboso (Cáceres)	*carcabosoentransicion.wordpress.com*
Cardedeu en Transició	Cardedeu, Barcelona	*cardedeuentransicio.wordpress.com*
Centro de Sostenibilidad de Aranjuez en Transición (CSA)	Aranjuez (Madrid)	*csaranjuez.wordpress.com*
Cercedilla en Transición	Cercedilla (Madrid)	
Cigales en Transición	Cigales (Valladolid)	*cigalesentransicion.wordpress.com*
Coín en Transición	Coín (Málaga)	*www.coinentransicion.com*
El Casar en Transición	El Casar (Guadalajara)	*www.elcasarentransicion.wordpress.com*
Es Castell en Transició	Menorca	
Es Mercadal en Transició	Mercadal (Menorca)	*www.facebook.com/esmercadal.entransicio*
Es Migjorn en Transició	Migjorn (Menorca)	*migjornentransicio.blogspot.com.es*
Fuengirola-Mijas en Transición (FMT)	Fuengirola/Mijas (Málaga)	*fuengirolamijastransicion.blogspot.com.es*
Gasteiz en Transición	Vitoria-Gasteiz (País Vasco)	*gasteizentransicion.wordpress.com*
Gran Canaria en transición	Gran Canaria	*www.facebook.com/grancanaria.entransicion*
Granada en Transición	Granada	*www.granadaentransicion.wordpress.com*
Granollers en Transició	Granollers, Barcelona	*granollersentransicio.wordpress.com*
Ibiza Isla en Transición	Ibiza (Ibiza, Baleares)	*www.ibiza-isla-transicion.com*
Instituto de Transición Rompe el Círculo	Móstoles (Madrid)	*www.mostolessinpetroleo.blogspot.com*
Intransition Marbella	Marbella (Málaga)	*www.intransitionmarbella.org*
Jerez en Transición	Jerez de la Frontera, (Cádiz)	*jerezentransicion.blogspot.com.es*
La Palma en Transición	La Palma (Santa Cruz de Tenerife)	*sites.google.com/site/lapalmatransicion*
La Puebla de los Infantes en Transición	La Puebla de los Infantes (Sevilla)	*lapuebladelosinfantesentransicion. wordpress.com*
Lanzarote en transición	Lanzarote (Las Palmas de Gran Canaria)	*www.facebook.com/lanzaroteentransicion*
Logroño en transición	Logroño (La Rioja)	*logronoentransicion.wordpress.com*
Málaga en Transición	Málaga	*www.facebook.com/pages/ Málaga-en-Transición/323452214342579*
Mancor Desperta	Mancor (Mallorca)	*sites.google.com/site/aramancordesperta*
Maó en Transició	Maó (Menorca)	*maoentransicio.org*
Red Menorca en transición	Menorca	*menorcaentransicio.org*
OSEL Transition Town	Novelda (Alicante)	*www.facebook.com/pages/ OSEL-Transition-town/182434118464996*
Pla Energétic Participatiu (PEP)	Barrios de Sant Martí, La Verneda y la Pau - Barcelona	*plaenergiaparticipatiu.cat*
Portillo en Transición	Portillo y Aldea de S. Miguel (Valladolid)	*portilloentransicion.wordpress.com*
Quijorna en Transición	Quijorna (Madrid)	*quijornaentransicion.blogspot.com.es*
San Martí - La Verneda en Transició	Barrio de Sant Martí - La Verneda (Barcelona)	*santmartilavernedaentransicio.wordpress.com*
Santa Coloma en Transició	Santa Coloma de Queralt (Tarragona)	*www.santacolomaentransicio.blogspot.com.es*
SEPA en Transición	Grupo de investigación. U. de Santiago de Compostela	

SPANISH TRANSITION INITIATIVES		
Name	Location	Web page
Sureste Ibérico en Transición	Córdoba-Almería-Murcia	matrizcelular.blogspot.com.es
Tarifa en Transición	Tarifa	transiciontarifa.foroactivo.com
Torremolinos en Transición	Torremolinos (Málaga)	torremolinosentransicion.blogspot.com.es
Tous en Transició	Sant Martí de Tous (Barcelona)	tousentransicio.blogspot.com.es
Transició VNG	Vilanova i la Geltrú (Barcelona)	www.transiciovng.blogspot.com.es
USC en Transición	Santiago de Compostela (La Coruña)	www.usc.es/entransicion
Valdepiélagos en Transición	Valdepiélagos (Madrid)	valdepielagostransicion.noblogs.org
Valladolid en Transición	Valladolid	valladolidentransicion.wordpress.com
Vallés en Transició	El Vallés (Barcelona)	vallesentransicio.blogspot.com.es
Zarzalejo en Transición	Zarzalejo (Madrid)	zarzalejoentransicion.blogspot.com.es
Zuera en Transición	Zaragoza	zueraentransicion.blogspot.com.es
Zurbarán en Transición	Bilbao (Vizcaya)	zurbaranentransicion.blogspot.com.es

It is interesting to notice that the formation of Transition initiatives accelerated over the time period covered by the research. Few of those documented (11 percent) were created before 2010, 47 percent arose in 2011 and the remaining 42 percent in 2012. However, only 18 main groups responded to our request for interviews about their dynamics, projects and so on. We assume those who did not respond are either in very early stages of internal organization and therefore unable to answer questions about their structure or activities, or not active. The further elaborations on groups' structures, areas of work, barriers and opportunities are therefore based on interviews with these 18 responding groups, which we assume are the more active initiatives.

Table 2.2.2 – Spanish Community-led Initiatives Similar to Transition, their Locations and Web Addresses.

SUMMARY OF COMMUNITY-LED INITIATIVES SIMILAR TO TRANSITION INITIATIVES IN SPAIN			
Movement or initiative	Examples	Location	Web page
De-growth	Desazkundea	País Vasco	web.desazkundea.org
	Decrece Madrid	Madrid	sindominio.net/wp/decrecimientomadrid
	Red de Decrecimiento de Sevilla	Sevilla	www.sevilladecrece.net
	Other examples include: Cataluña, Navarra, Aragón, La Rioja, Asturias, Canarias		www.decrecimiento.info
Integral Cooperatives	Cooperativa Integral Catalana (CIC)	Cataluña	cooperativa.cat/es. Specific projects include calafou.org and www.aureasocial.org/es
	Cooperativas Integrales Oeste Norte (CION)	North East of Iberian Peninsula	www.cooperativasintegraleson.net
	Cooperativa Integral Aragonesa (CIAR)	Aragón	ciar.cc
	Other examples include: País Vasco, Ibiza, Albacete, Granada, Valencia, La Rioja, Madrid, Zamora		integrajkooperativoj.net

SUMMARY OF COMMUNITY-LED INITIATIVES SIMILAR TO TRANSITION INITIATIVES IN SPAIN			
Movement or initiative	Examples	Location	Web page
Ecovillages	Lakabe	Navarra	*rie.ecovillage.org*
	Matavenero	Castilla y León	
	Los Portales	Sevilla	*www.losportales.net*
	Sunseed	Almería	*www.sunseed.org.uk/es*
	Molino de Guadalmesí	Tarifa	*www.molinodeguadalmesi.com*
	Valdepiélagos	Madrid	*www.ecoaldeavaldepielagos.org*
	Many other examples		*rie.ecovillage.org*
Permaculture	Permacultura Monsant	Tarragona	*www.permacultura-montsant.org*
	Red de Permacultura del Sureste (REPESEI)	Sureste Ibérico	*www.permaculturasureste.org*
	Permacultura Mediterránea	Mallorca	*www.facebook.com/PermaMed/*
	Proyecto Pachamama	La Palma	*proyectopachamama.blogspot.com.es*
	Permacultura Barcelona	Barcelona	*www.facebook.com/PermaculturaBarcelona*
	Permacultura Cantabria	Cantabria	*www.permaculturacantabria.com*
	Iraun Permakultura	Gasteiz (País Vasco)	*iraunpermakultura.wordpress.com*
	Many other examples		
Other initiatives	Slow Movement	Barcelona, elsewhere	*movimientoslow.com*
	Economía del Bien Común	Madrid, Barcelona, elsewhere	*economia-del-bien-comun.org/es*
	Red Sostenible y Creativa	Valencia	*www.sostenibleycreativa.org*
	Poc a Poc	Mallorca	*www.pocapoc.org*
	Vespera de Nada	Galicia	*www.vesperadenada.org*
	Urban garden networks	Madrid	*redhuertosurbanosmadrid.wordpress.com*
		Barcelona	*huertosurbanosbarcelona.wordpress.com*
	Interchange networks: time Banks and local currencies	Whole Country	*mapa.vivirsinempleo.org/map*
	Consumer groups	Madrid	*gruposdeconsumo.blogspot.com.es*
		Cataluña	repera.wordpress.com
	Social Centres	Whole Country	Can Masdeu, Barcelona. *www.canmasdeu.net/es*
			La Tabacalera, Madrid. *latabacalera.net*

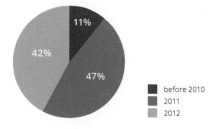

Figure 2.2.3 – Annual Rate of Creation of New TIs in Spain.

2.2.3. On What and How do Spanish TIs Work? Assessing Difficulties and Opportunities

As shown in the figure below, we asked TIs about the numbers of working groups within each of them, which topics they cover, how many projects/activities they are carrying out, and the status of each project. Figure 2.2.4A (the left-hand diagram in Figure 2.2.4) gives an initial overview of the groups' activities. More than half of the 18 initiatives interviewed have between five and ten active working groups. A single initiative (comprising the five percent slice) has more than ten active working groups. Of the remaining TIs, three (16 percent) have fewer than five working groups and two (11 percent) have no defined working groups at all.

As a prelude to a more qualitative assessment, Figure 2.2.4B gives a quantitative overview of the status of these activities (whether projects have been completed, are under implementation, or are still in a planning phase). It is noteworthy that more than half of the activities reported have already ended, and 35 percent are being actively implemented; only 9 percent are new activities still at the inception phase. This means the reported data on activities do not refer to aspirations – ideas or future projects – but reflect how the activities and projects of the majority of participating TIs have actually developed.

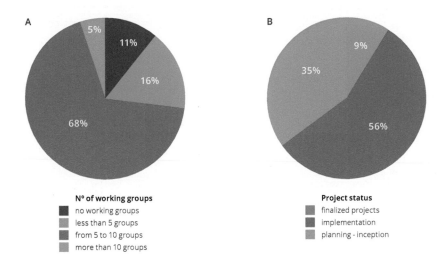

Figure 2.2.4 – *Numbers of Working Groups and Status of Projects in Responding TIs.*

The 18 TIs reported a total of more than 40 projects that they are developing (or have developed. The most popular topics are education and food production, with the greatest number (40 percent) of projects based on education: dissemination of transition and sustainability principles, and behavioural and personal change. According to groups' reports of the outcomes of their activities, these are also the most successful projects. Food is the second most common project topic: 30 percent of reported projects focused on local/organic food production. Food projects seem to be among the most straightforward way for a group to take action beyond its first educational projects, and are also reported as generally successful. A different picture is reported for projects with a more practical orientation in areas other than food, such as creating an energy plan, pushing for a local currency, or energy production. We didn't measure the success or failure rates of projects/activities as such, or seek to identify indicators of success or failure. We were interested in understanding which kinds of projects or actions were the most difficult or easy to undertake, in order to understand the barriers and opportunities that Spanish TIs encounter.

Examples of projects

Among many different types of projects developed by TIs, those based on education and local food were clearly reported to be the most numerous and feasible, while those on energy transition, local money, co-housing or use of public spaces seem to be more challenging to take on. This section moves on to describe various examples identified during the research. They have been organised in four themes: Food, Education and Awareness-raising, Energy, and Economy.

Food

Most initiatives have a project related with food. The most numerous such projects are collective urban gardens, like *Butroi en Transición*, *Universidad de Santiago en Transición* and *La Puebla de los Infantes en Transición*. Also very common are consumer groups that connect local producers with a self-organised group of consumers, examples of which are operated by *Cardedeu en Transició* and *Granollers en Transició*. Other projects include community supported agriculture projects (CSA), like *Zarzalejo en Transición*, communal henhouses, like in *Granada en Transición*, or local markets, like in *Coín en Transición* or *Santa Coloma en Transición*.

Figure 2.2.5 *– CSA from Zarzalejo en Transición. Credit: Zarzalejo en Transición.*

Education and Awareness-Raising

Almost all initiatives start with small awareness-raising projects in their communities: workshops, talks or meetings to explain the main issues we are facing as a society and the methodology of the Transition movement. A few examples are the Berenars de Transició of *Cardedeu en Transició*, or the videoforums done by *Albacete en Transición*. They also include educational projects in schools like in *INtransition Marbella*, or courses in universities like *Universidad de Santiago de Compostela (USC)*.

Figure 2.2.6 – *Berenar de Transició of Cardedeu en Transició. Credit: Cardedeu en Transició.*

Energy

Less numerous are projects related directly with energy. We have projects for insulating homes, like *Zarzalejo en Transición's* home insultation program, supported by the local council. Projects to install solar energy on roofs of community buildings include *Pla de Energia Participatiu* in the Sant Martí i la Verneda neighborhood in Barcelona. There are also related projects on local transport, such as promotion of carsharing by *Lanzarote en Transición*.

Economy

Several projects address local economies and exchange systems. We find local currencies like El Zoquito from *Jerez en Transición* or la Turuta from *Vilanova i la Geltrú en Transició*. There are also time banks like that of *Cardedeu en Transició* and barter markets like the MIM of *Menorca en Transició*.

Figure 2.2.7 – *Solar Energy Project in Communitary Roofs from Pla de Energia Participatiu (PEP) in the Sant Martí i la Verneda Neighborhood in Barcelona. Credit: PEP.*

Figure 2.2.8 – *Interchange Market of Menorca (MIM). Credit: Menorca en Transició.*

2.2.4. Barriers and Opportunities for Transition Initiatives

Considering challenges, we asked TIs about the main difficulties and barriers they experience executing their projects. From the results, three main themes emerged: lack of financial support and funding sources, difficulties of getting more people involved in the projects and so guaranteeing continuation of the proposed actions, and finally a general lack of communication among group members and among different initiatives. The following paragraphs address each of these in turn.

With the exception, as mentioned in the previous section, of largely self-resourced projects on education and food production, a major perceived challenge for all participating TIs is the lack of financial support and funding sources. As chart A in Figure 2.2.9 shows, the majority of projects (82 percent) relied solely on TIs' own resources, while a few (12 percent) obtained some public funding, and an even smaller number (six percent) received private financial support.

> *While financial and other external support is often a key factor for success, at the core of the Transition movement is the capacity for self-organization as a new, more sustainable and resilient pathway for community development.*

However, financial support is not – or should not be – the main reason for success or failure of a Transition project: many other, non-financial forms of support are essential for Transition activity. Chart B in Figure 2.2.9 hence shows what proportion of Transition projects have received local government support other than funding. This includes access to space, advertising and promotion, awarding licenses for activities, and others. The most obvious cases of support are those of Mijas, in which the municipality has made public spaces available at no cost for urban agriculture projects managed by the transition initiative Mijas en Transición, and Marbella, in which the municipality donated use of the city council hall as the venue for a Regional Transition meeting in October 2012. Aside from these clear examples of fruitful collaboration, just one third of the projects developed by TIs studied in the survey had received some kind of non-financial support from public institutions. The majority (61 percent) have never received any support. A further ten percent of reported examples are projects in preparatory or planning phases in which public institutions have been contacted but it is not clear whether their support will ultimately be forthcoming.

While financial and other external support is often a key factor for success, at the core of the Transition movement is the capacity for self-organization as a new, more sustainable and resilient pathway for community development. Related movements like degrowth, for example, agree with the latter point and indeed reject reliance on external and financial support, seeing the source of revolutionary power for a transition in the process of self-organization itself, and the breaking with previous dependencies and institutional rules this entails. In this light, concerns with external support to TIs projects would be less relevant than they appeared during the interviews. Accordingly, we tried to focus more on the internal dynamics of Transition groups and associated coordination issues, corresponding to the other two main emerging themes on challenges: involving people and communication.

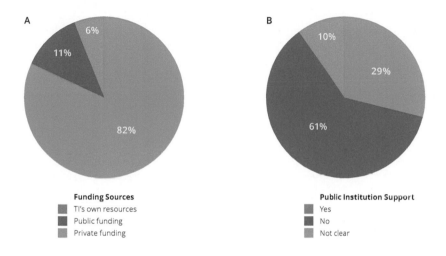

Figure 2.2.9 – *Transition Projects Receiving Financial and In-kind Support.*

The interviews with all 18 TIs showed an almost complete consensus about the challenge of ensuring group members' involvement in activities and sustaining this in the medium and longer term. Lack of participation and difficulties of internal coordination in groups were raised in almost all interviews. Factors affecting both participation in activities of existing members and the recruitment and integration of new members include: lack of individual motivation for longer term involvement, lack of financial support to guarantee continuity of members' involvement in specific tasks, lack of specialist knowledge of how to manage groups and networks, and a scarcity of strategic planning for development of the TI (activities, groups, networks). A general characteristic of the Spanish context that became apparent, and related to such difficulties, resides in an attitude unfamiliar with either planning or closely following long term plans, instead being extremely adaptable to changing conditions and roles. In such an environment, in general, the social impacts of TIs is great, with high attendance at organized events, but moving on from this to motivate people to active involvement in activities was reported to be very challenging. However, when stating those observations, we need to take into account the very recent and emergent nature of the Transition movement as a Spanish phenomenon, and the experience and time necessary to create a dynamic appropriate to the context in order for things to take off.

2.2.5. Working Across Scales: Navigating from Personal Resilience to Societal Transformation

Moving on from challenges to opportunities, we first consider the situation regarding communication and networking strategies in Spain, which naturally leads to a consideration of how an embryonic movement develops the capacity to work effectively across scales. Previous sections have mentioned how communication within and among groups, TIs, and different movements seems to lack intensity and consistency. As an indication, no TIs interviewed mentioned cross-group collaborations or cross-movement projects. It is also notable that while all the TIs self-identify as 'Transition Initiatives' because they like and agree with the principles and practices of the Transition movement, 80 percent of them were not registered with or in touch with Transition Network,[28] the international coordinating and support organisation for the Transition movement. These initiatives prioritize internal issues over building links with international networks, and do not actually see the international network as an important source of resources (mainly because materials are mostly in English and not adapted to the local context). However, all of the TIs participating in the project recognized the need to build a Spanish network in order to foster collaboration and communication nationally; sharing resources, experiences and plans.

On this basis, the national Transition hub (*Red de Transición España, REDTE*) and website were launched in early 2014 to act on the consensus among initiatives about the value of coordi-nation at national level. Members of the national hub also believe that creation of a national network is an essential preliminary step to creating the necessary links with international ini-tiatives and networks, sharing the specific Spanish cultural experience with other movements and initiatives. While few Spanish TIs are currently registered with the international Transition Network, they all believe in the power of Transition as a movement and the value of openness to involvement of any other interested member or group. The willingness to strengthen the national network in a country with high internal cultural diversity and regional political auton-omy seems a very positive signal. It indicates that during its current phase of 'exploitation' (in the terms of resilience theory; rapid initial growth in non-technical language), the Spanish Transition movement has the motivation and ambition to build on its momentum and upscale from various local experiments towards much wider-reaching social change.

This clarifies some subtle aspects of the relationships among group resilience, empowerment of individual members, and a group's capacity to promote wider Transition, and how this unfolds over time. At a group's inception, resilience represents the capacity to learn and experience how to build and improve personal, and consequently collective, capacities: firstly, to become more aware of our own dependencies (functions of our lifestyles and habits), and secondly, as a consequence of this, to become a little more self-sufficient through direct engagement with production of primary goods such as food or energy. In this way, the group becomes a context that supports the growth of personal resilience – in terms of the capacity to embrace change in lifestyle, priorities and habits – among its members. This is something

28   *http://www.transitionnetwork.org*

we detected in every group we interviewed, and shows the importance of the proliferation of new groups even if at this stage their concrete external achievements appear minor or non-existent. Experiencing resilience in this way is something that has little to do with conscious planning: it is about engaging people at a much more emotional level.

Moving from experiencing resilience at this very local and even personal level to seeking to effect change at higher levels (larger spatial scales and/or longer time scales) is a key step towards broader societal transitions, and a major challenge in the life of any Transition initiative. Acceptance of the need for change at community or societal scale, and commitment to effecting such change, is a very different process requiring far greater emphasis on planning, cooperation and engagement. The first stage is consistent with the notion of resilience as adapting to change in order to maintain continuity of existing system function, as the personal changes involved don't necessarily rely on prior changes at higher levels. The second implies far-reaching structural and functional transformation to entirely new configurations, most likely involving radical shifts in general outlook and mindsets.

> *(...) the group becomes a context that supports the growth of personal resilience – in terms of the capacity to embrace change in lifestyle, priorities and habits – among its members.*

As our brief assessment shows, this step from human scale resilience to systemic societal transformation is the key issue both for individual Transition initiatives in Spain, and for the Spanish network as a whole. To make it happen requires engagement, networking, coordinated planning and collective strategic action: the main issues raised when Spanish TIs reflect on the challenges they face. Simultaneously, enormous potential for such upscaling exists in the aftermath of the economic crisis of 2008. The dramatic and unavoidable reshaping of lifestyles and changes in quality of life that have persisted in Spain ever since may increase receptivity among the wider public to the efforts of Spanish TIs to foster the scale of engagement and action necessary for wider transition.

This assessment was a first tentative attempt to gather data on the scale, challenges and opportunities in Spain for transition towards more resilient and sustainable communities. Further investigation will be necessary to identify the cultural, political and strategic factors that can enable the shift from mainly local action to wider societal transition. This will need to involve close dialogue with changes in practice, both for theory to inform practice and for hands-on experience to feed back into the reassessment of theory of understanding and manage societal transitions towards sustainability and resilience.

2.3. Resilience and Community Action in Bristol

TOM HENFREY

Bristol has a reputation, deservedly or not, as one of the world's greenest and most resilient cities. It was chosen as European Green Capital for 2015, and among the first 33 cities worldwide in the Rockefeller Foundation's 100 Resilient Cities programme. These accolades in large measure build upon the achievements of the city's dynamic and vibrant movement for community-based action on sustainability. Impressive

> *Bristol has a long and proud tradition of grassroots environmental action.*

as these achievements are, genuine progress is minor relative to the continued scale and momentum of the fossil-fuel based economy. This chapter examines the recent history of grassroots environmentalism in Bristol through a resilience lens, focusing on the nature and consequences of cross-scale interactions involving and affecting community level action.

Bristol has a long and proud tradition of grassroots environmental action. Emmelie Brownlee's green history of the city dates its origins to the late 1960s, when a coalition of community groups successfully opposed construction of an inner ring road that would have dramatically altered the urban landscape, with the local Friends of the Earth Group the first of many pioneering environmental initiatives launched during the 1970s.[29] Precedents go back further, at least to the founding in 1788 of a group dedicated to the abolition of slavery, the first outside London. Historian Steve Hunt sees the city's garden suburb movement, inspired by Ebeneezer Howard's Garden Cities movement in the early decades of the twentieth century but later undermined by drives towards low-cost housing, the pragmatic demands of post-war urban reconstruction, and

29 Brownlee, E., 2011. *Bristol's Green Roots: the growth of the environmental movement in the city.* Bristol: The Schumacher Centre.

the reshaping of cities to accommodate private cars, as anticipating many of the present-day ideas of the Transition and permaculture movements in its concerns with integrated working and residential neighbourhoods, urban environments conducive to residents' mental and physical health, ready access to green spaces, attention to social capital, desire to foster community cohesion through provision of social amenities, emphasis on community ownership of housing, association with cooperative enterprise and attention to edible planting.[30]

Active since 2007, Transition Bristol is tenth on Transition Network's list of 'official' initiatives, making it the world's first Transition city. Like many big-city initiatives, it quickly shifted the emphasis to smaller neighbourhood groups focused on local projects, with the city group taking more of a strategic and coordinating function.[31] More recent years have seen the emergence of several city-wide initiatives, in energy,[32] food, education,[33] well-being,[34] transport,[35] and with the establishment of the Bristol Pound as a city-wide complementary currency, whose ten-pound note, fittingly, bears the image of abolitionist Hannah More.[36] Some of these arose from neighbourhood groups, some directly from within Transition Bristol, others entirely independently but based on common aims and values. Transition Bristol directly initiates very few of the events and activities listed in its website and newsletter[37] or posted on its Facebook group. Its core team members describe Transition in Bristol as this broader movement and network rather than the organisation itself.

This wealth of activity can create a compelling illusion that Bristol is a truly sustainable and resilient post-carbon city. The cycling infrastructure, while it appears shockingly primitive to visitors from Germany and the Netherlands, is as good as I have seen in any British city.[38] Thriving networks of allotments and community gardens support exchange of skills, knowledge, plants and other resources among growers, and larger projects on the outskirts of the city provide a range of options for buying local organic produce. Numerous community energy projects – including two separate renewable energy co-operatives – collaborate as the Bristol Energy Network, which produced the first Community Energy Strategy for any British city in 2013,[39] a few months before the UK government released its national equivalent. At the time of writing in December 2014, nearly 700 independent small businesses in the city

30   Hunt, S., 2009. *Yesterday's Tomorrow: Bristol's Garden Suburbs*. Bristol Radical Pamphleteer #8. Bristol Radical History Group.

31   *http://transitionbristol.net/transition-bristol-timeline/*. Accessed Jan 2nd 2015.

32   *http://www.bristolenergynetwork.org/*. Accessed Jan 2nd 2015.

33   *http://www.shiftbristol.org.uk/*. Accessed Jan 2nd 2015.

34   *http://www.happycity.org.uk/*. Accessed Jan 2nd 2015.

35   *http://www.livingheart.org.uk/*. Accessed Jan 2nd 2015.

36   *http://bristolpound.org/*. Accessed Jan 2nd 2015.

37   *http://transitionbristol.net/category/events/*. Accessed Jan 2nd 2015.

38   The Bristol Cycling Manifesto sets out a strategy for dramatic improvement of infrastructure in the city at *http://bristolcyclingmanifesto.org.uk/manifesto/*. Accessed Jan 2nd 2015.

39   *http://bristolenergynetwork.org/strategy*. Accessed Jan 2nd 2015.

Figure 2.3.1 – Allotment in Bristol. Credit: Gesa Maschkowski.

accept the Bristol Pound, whose Real Economy Project seeks to use the currency as a tool to develop an alternative economy, localised and socially responsible.[40] Nearly 1000 people subscribe to the Bristol Permaculture Group's email list, with over 50 more people every year training on the city's annual permaculture design certificate course or the more advanced Practical Sustainability Training offered by permaculture teaching co-op Shift Bristol. It's easy to move through the city and visit only places and meet only people who would support this impression: a comfortable microcosm of a society that has already made the Transition away from dependence on fossil fuels and economic growth. Some guide books detail walking and cycling routes that show how to do exactly that.[41]

Grassroots activism also benefits from a number of cross-sector partnerships, in which business and local government enable and directly support the creation of common pool resources through grassroots action, and actively work towards their upscaling. These include organisations that emerged from the grassroots and have grown to become part of the establishment. Sustrans, Britain's national cycling charity and custodian of long-range national

40   *http://bristolpound.org/real-economy.* Accessed Jan 2nd 2015.
41   Walmsley, E. (ed.), 2010. *Bristol: A Guide to Good Living.* Bristol: Alastair Sawday Publishing.

cycle routes, began life as a local pressure group and continues to extend and maintain cycle routes in the city. A former project manager at Sustrans set up Roll for the Soul, a cycling cafe and networking and event space in central Bristol operated as a Community Interest Company. The Centre for Sustainable Energy grew from its roots as a volunteer-run experiment in retrofitting derelict urban properties in the 1970s to chief delivery organisation for national government programmes in energy efficiency in the 2000s. It continues to maintain a careful and sensitive presence in Bristol's thriving community energy scheme, supporting – but not leading – the projects and process that led to the creation of the Bristol Energy Network, and, just as important, keeping out of the way when its support is not needed. The Bristol Pound's financial services are delivered via a partnership with the local Credit Union, a member-owned provider of financial services.

Figure 2.3.2 – *Urban Gardening, Bristol. Credit: Gesa Maschkowski.*

Much grassroots action also benefits from direct support from local government. City Council decisions to accept payment of council tax and business rates in Bristol Pounds greatly strengthen the currency's viability.[42] The Bristol Food Policy Council,[43] a council-run initiative supported by EU funding, works closely with the Bristol Food Network, an association of businesses and community projects,[44] on developing and implementing the Good Food Plan for Bristol.[45] The Good Food Plan built on an earlier report, Who Feeds Bristol, commissioned by the City Council and NHS, and something of a successor to the Bristol Peak Oil Report,[46] produced by Transition Bristol with financial support

> *Grassroots activism also benefits from a number of cross-sector partnerships, in which business and local government enable and directly support the creation of common pool resources through grassroots action, and actively work towards their upscaling.*

from the City Council. Two of the city's key community food growing projects – the CSA at Sims Hill Shared Harvest[47] and the neighbouring education and demonstration site Feed Bristol, run by the Avon Wildlife Trust[48] – operate on prime agricultural land provided rent-free by the City Council.[49] All of these grassroots initiatives and associated collaborations featured prominently in Bristol's successful applications to become European Green Capital – itself the outcome of a long-term, and ongoing, cross-sector partnership - and for participation in the Rockefeller Resilient Cities Programme. In all these instances – and many others – cross-scale interactions are of a nature we would expect to have positive benefits for local economic resilience: community-scale innovation beyond the limits of what local government can achieve directly, both benefits from public sector support and nurturance, and feeds into larger-scale initiatives, both directly within the City Council and through national and international initiatives in which it is involved, with many of the resulting initiatives being delivered as or in collaboration with local businesses, or taken up by them.

---

42   http://bristolpound.org/news?id=38. Accessed Jan 2nd 2015.

43   http://bristolfoodpolicycouncil.org/. Accessed Jan 2nd 2015.

44   http://www.bristolfoodnetwork.org/about/. Accessed Jan 2nd 2015.

45   http://bristolfoodpolicycouncil.org/wp-content/uploads/2013/03/Bristol-Good-Food-Plan_lowres.pdf.
     Accessed Jan 2nd 2015.

46   http://www.20splentyforus.org.uk/UsefulReports/Bristol_Peak_Oil_Report.pdf. Accessed Jan 2nd 2015.

47   http://simshill.co.uk/. Accessed Jan 2nd 2015.

48   http://www.avonwildlifetrust.org.uk/feedbristol. Accessed Jan 2nd 2015.

49   http://simshill.co.uk/2010/12/01/sims-hill-in-recent-bristol-city-council-press-release/. Accessed Jan 2nd 2015.

Despite these achievements, Bristol remains, overwhelmingly, a city dependent on the global fossil fuel economy, with all its negative consequences for sustainability, ethics, equity and resilience. The Who Feeds Bristol report estimated that 84 percent of food retail purchases take place in outlets of the five major national supermarket chains – higher than the national average – and suggests that the numbers of such stores per head of population are higher than in other major UK cities.[50] The two largest companies in greater Bristol – and among its biggest employers - are Imperial Tobacco, whose annual turnover is several billion pounds and which sells tobacco products in 160 countries worldwide, and Airbus, a plane manufacturer whose sales portfolio includes military clients.[51] Levels of congestion on the city's roads every weekday morning and afternoon clearly show that transport policy is dominated not by concerns with mobility, but with cultivating captive markets for asphalt, cars and people.

All of these grassroots initiatives and associated collaborations featured prominently in Bristol's successful applications to become European Green Capital.

Bristol's claims to be among the greenest and most resilient cities in the UK may be well founded: but this casts a harsh light of realism on just how remote resilience remains as a goal.

Many examples where lock-in to existing systems is hampering transitions to resilience indicate the power of the 'Remember' effect. A recent history of the green movement in Bristol concluded that fragmentation among existing groups – sometimes associated with outright competition for scarce funding and other resources – means it is, in total, rather less than the sum of its parts.[52] The launch of the Community Energy Strategy in June 2013 appeared to be rather co-opted by the announcement of a new programme led by the unreflectively business-as-usual West of England Local Enterprise Partnership, Bristol Solar City, whose stated ambition was to install a gigawatt of new photovoltaic output by 2020 but which subsequently vanished, virtually without trace.[53] Meanwhile, central government plans move forward to construct two new industrial biomass power plants and a 3.2 gigawatt nuclear power station within the vicinity of the city, in the face of strong local opposition.[54]

50   Carey, J., 2011. *Who Feeds Bristol: a baseline study of the food system that serves Bristol and the Bristol city region.* Pp. 22-24. *http://www.bristol.gov.uk/sites/default/files/documents/environment/environmental_health/Who-feeds-Bristol-report.pdf.* Accessed Jan 2nd 2015.

51   Bache, R. 2014. The £40 Billion Club. Pp. 8 in The Business Guide 2014. Western Daily Press. *http://www.southwestbusiness.co.uk/southwestbusiness-co-uk/_img/Supplements/Top_150_Businesses_Guide_2014..pdf.* Accessed Jan 2nd 2015.

52   Brownlee, *op. cit.*

53   *https://www.facebook.com/bristolsolarcity/*

54   *http://www.avoncoalitionagainstbigbiofuels.org.uk/*
      http://www.newstatesman.com/politics/2014/10/hinkley-nuclear-power-plant-bombshell-out-going-european-commission Accessed Jan 2nd 2015.

Perhaps most ironically of all, the City Council has approved construction of a new road that will cross land it currently leases to the Feed Bristol project, destroying some of the city's very limited area of prime agricultural land along with at least 25 other green and public spaces.[55]

The tension between community action on resilience and inertia in incumbent systems is not just circumstantial or contextual, but manifests directly in its interactions with higher-level strategic programmes, including Green Capital and 100 Resilient Cities. Each, in a different way, could be seen as to some extent appropriating the achievements of grassroots actions by top-down initiatives that deploy key terms like resilience and sustainability in more conservative ways. In cross-scale interactions of this type, institutional memory at high levels restricts the potential for small-scale innovations to escalate up the panarchy in revolutionary ways (in other words, the 'Remember' effect dominates the 'Revolt' effect).

Figure 2.3.3 – Proud to accept Bristol Pounds. Credit: Gesa Maschkowski.

Bristol's Green Capital bid was the outcome of a careful process of partnership-building, loosely held by the council but involving hundreds of organisations and groups from across the city and anticipated to be an ongoing endeavour far beyond 2015 itself.[56] Concerns at transfer of many implementation responsibilities to a new company, Bristol 2015, whose main function appeared to be to attract private sector sponsorship and inward investment for the Green Capital programme, turned to consternation at a general lack of inclusion and transparency, leading to removal of the company's highly-paid chief executive just three months before the start of 2015 itself.[57] Whether this amounts to the replacement of a 'remember' effect with a 'revolt' effect – in other words, a return to the partnership as the core framework for green capital activities, with the company in a support role - remains to be seen.

The main stated aim of the Rockefeller programme is the creation of a learning network among 'Chief Resilience Officers': high-ranking local government officials in each of the cities involved,

55   *https://thebristolcable.org/2015/02/metrobus-a-timetable-of-destruction-for-bristols-green-capital/.* Accessed Dec 18th 2016.

56   *http://bristolgreencapital.org/.* Accessed Jan 2nd 2015.

57   *http://www.bbc.co.uk/news/uk-england-bristol-29472188.* Accessed Jan 2nd 2015.

each of whom will oversee the development of a resilience plan with the support of Rockefeller and commercial partners.[58] The impression is of a clear hierarchy of top-down influence, with grassroots involvement presumably at the discretion of participating authorities and clearly subordinate to the centralised processes through which resilience is defined and operationalised in the programme. A Resilience Action Group has been set up within the more flexible and participatory framework of the Green Capital Partnership. Its remit includes providing a platform for grassroots engagement with the Rockefeller Programme and supporting the Chief Resilience Officer in cross-scale aspects of their work.[59]

Initial conversations among members of the Resilience Action Group indicated a range of perspectives on what resilience is and ideas of how to achieve this. These converged on more radical notions of resilience, with an emphasis on ongoing systemic change, both incremental and transformative, rather than resisting change or 'bouncing back'. Many members of the group talked about the need for ongoing learning processes to promote flexible and proactive responses to change and build adaptive capacity, and the need for this to involve all sectors of

Inclusion in planning and decision-making was a key factor, raised in many different contexts and as both a practical and an ethical issue: in other words so that adaptive capacity can draw upon the greatest range of perspectives and knowledge, and that resilience can be for the benefit of all.

society. Inclusion in planning and decision-making was a key factor, raised in many different contexts and as both a practical and an ethical issue: in other words so that adaptive capacity can draw upon the greatest range of perspectives and knowledge, and that resilience can be for the benefit of all. In summary, Mike Zeidler of Happy City suggested that this reflects a recognition that cities are human systems, and resilience in cities involves a human response that can recognise the need for change and understand how to bring it about. Revolt – in the form of shifts to more inclusive, human-centred forms of decision-making via the upscaling of learning and transformational processes already underway at small scales and consequent reconfiguration of higher level systems and processes – is a clear implication.

As the foregoing account suggests, any such transformation will need to combine building

58   http://www.100resilientcities.org/about-us#/-_/. Accessed Jan 2nd 2015.
59   http://bristolgreencapital.org/resilience-action-group/. Accessed Jan 2nd 2015.

Figure 2.3.4 – *Mural Painting Nuclear Waste, Bristol. Credit: Gesa Maschkowski.*

on existing areas of cooperation and support with directly opposing undesirable initiatives of both local and central government. The Avon Coalition Against Big Biofuels mobilised enough public support to force a public hearing about a planning application for a proposed biomass installation in Avonmouth, which councillors rejected due to concerns over air quality and attendant public health risks.[60] Local campaign group Frack Free Bristol raised a petition of over 5000 signatures in favour of a ban on fracking in the city, provoking formation by the City Council of a cross party working group to discuss the proposal.[61] Members of the Blue Finger Alliance seek to secure 1000 hectares of prime agricultural land on Bristol's urban fringe for food production and associated employment, training and amenity opportunities.[62] The Blue Finger land includes the council-owned smallholdings where the Sims Hill and Feed Bristol projects are located, threatened by the controversial Metrobus project. The Alliance's vision seeks to make this situation an opportunity to initiate a high level conversation involving the local councils about how best to use this valuable resource, including via partnership working and joined-up decision making.[63] Many of these campaigns draw attention to the particular

60   *http://www.avoncoalitionagainstbigbiofuels.org.uk/balfour-beattynexterra-11mwe-biomass-gasifier-rejected-by-bristol-city-council/.* Accessed Jan 2nd 2015.

61   *http://www.bristolpost.co.uk/Councillors-debate-fracking-allowed-Bristol/story-20822076-detail/story.html.* Accessed Jan 2nd 2015.

62   *http://www.bluefingeralliance.org.uk/blue-finger-explained/article-downloads/vision/.* Accessed Jan 2nd 2015.

63   Pugh, S., P. Summers & M. Longhurst, 2014. *The Blue Finger Vision. Business as usual? Or an opportunity to create a world class Urban Agriculture Hub? http://www.inuag.org/sites/default/files/The-Blue-Finger-Vision.-A-world-class-hub-of-urban-agriculture-for-the-Bristol-city-region.pdf.* Accessed Jan 2nd 2015.

Figure 2.3.5 – Bristol Mural Painting-Inclusive Community. Credit: Gesa Maschkowski.

folly of initiating such schemes during Green Capital year, when Bristol's claims to be a sustainable city have the attention of the entire world. It's possible that a shift in the balance from 'Remember' towards 'Revolt' could be the tipping point towards a powerful transformation.

A new proposal emerging from the Green Capital Partnership is for Bristol to become a One Planet City. Significantly upscaling community-level work undertaken over the past few years by grassroots organisation One Planet Bristol,[64] this would involve residents, community groups, businesses, and local authority working in partnership, supported as necessary by national government, to operationalise the concept of One Planet Living and create a long-term legacy of meaningful action beyond 2015 itself.[65] The transformative potential of such a vision – if implemented through the lens of Type 4 resilience (see page 94), rather than more limited notions of greening the economy – is clear. Although within tangible reach, it will face major challenges: whether community action for resilience in Bristol is now coming of age remains an open question.

64   *http://www.oneplanetbristol.com/.* Access Jan 2nd 2015.

65   *http://sustainablecitiescollective.com/david-thorpe/1020916/bristol-takes-steps-towards-becoming-uks-second-one-planet-city.* Accessed Jan 2nd 2015.

TRANSITION AND RESILIENCE

3

3.0. Linking Theory and Practice of Community Resilience
HELEN ROSS

Resilience – and its complement of achieving transformation to more desirable systems – is an appropriate goal in attempting to manage the complex problems of human and environmental change. However, like the Indian metaphor of blind men feeling different parts of an elephant and drawing confident but incorrect conclusions as to the nature of the object before them, there are several different strands of resilience thinking, from different disciplines or interdisciplinary fields, and there is little communication or sharing across these. Since one specialises in ecosystem behaviour (recognising society), another entirely in human dimensions, and a third in disasters (there is also 'engineering resilience'), these offer excellent prospects for reconciliation to provide us with insights and guidance for managing and where necessary changing our social-ecological systems. Community action can contribute to this reconciliation.

Figure 3.0.1 – Helen Ross. Credit: Gesa Maschkowski.

Each of these strands of resilience thinking is relevant to an understanding of 'community resilience': a term often used in both policy and activism, but which until recently has lacked a consistent grounding in resilience science. That of social-ecological systems is couched in the paradigm of complexity, a field that understands the world in terms of the behaviour of 'complex adaptive systems'. This rejects the idea of our world being governed by clear cause-and-effect relationships and predictable linear trends, to view it as consisting of far more complex sets of interacting patterns – just think of the variables and interactions involved in producing our daily weather. Despite being termed 'social-ecological', this is so far limited in the understanding and theorisation of social dimensions. A psychology and health science strand has built from studies of how some individuals thrive despite major adversities in their lives, to understand the nature of social resilience: what strengths do communities and societies hold, and can they be enhanced? Unfortunately this strand has tended to neglect environmental roles. Meanwhile, disaster resilience combines ideas from the psychology and mental health literature (but more focused on individuals and households), with engineering resilience, the need for infrastructure that can stand up to major natural disasters.

Understanding resilience in these multiple ways is highly important for community action. Taking the best aspects of each, I first advocate understanding our world in terms of complex adaptive systems, full of uncertainties, not as a predictable system that responds linearly to interventions, or behaves according to clear trends. I advocate thinking in terms of social-ecological systems (including the economic), but developing a far more sophisticated understanding of the many ways in which the human dimensions drive their ecosystems and adapt to change. Social change through activism, and understanding and addressing power relationships in the process, is a crucial but hitherto neglected area. Then, I advocate attention to the psychology and mental health focus on people's strengths rather than their weaknesses and vulnerabilities. Berkes and Ross (2013) listed a number that have been identified in research to date (Figure 3.0.2). They note that these are probably necessary, but not sufficient: agency and self-organising are needed to convert latent strengths into active processes when required.

Figure 3.0.2 – *Community Resilience as a Function of the Strengths or Characteristics that have been Identified as Important, Activated Through Agency and Self-organization (Berkes and Ross 2013).*

Figure 3.0.3 – *Repair Café, Transition Bonn. Credit: Gesa Maschkowski.*

The concept of adaptive capacity, having the appropriate strengths in place to adapt when required, overlaps strongly with the idea of resilience. While there are different ways of understanding this relationship because of differing views as to how much resilience is a process or an outcome, for convenience we can view adaptive capacity as the potential to cope or thrive through major challenges (or disturbances), and resilience – or transformation - the processes and outcomes of doing so. Thus community action could be focused on building adaptive capacity, focusing on the sets of strengths known to contribute to adaptive capacity, and communities' senses of agency and capability for self-organising.

> *Agency and self-organising are needed to convert latent strengths into active processes when required.*

Like the social-ecological systems theorists and activists in this volume, I advocate understanding resilience as a multi-level phenomenon, in which local links readily to global, and the resilience of individuals contributes to and receives from the resilience of their communities and places.

I thus see two major opportunities for community action. Community developers, taking a resilience perspective, have a clear role in building and integrating selected community strengths, and empowering communities to self-organise and exert agency in order to build the resilience of the local parts of the global system they steward. Community actors, and activists, can adopt a resilience perspective to reshape their social and ecological goals: from past and current goals towards improving the world and local places, to building the resilience of their parts of the globe (and global systems such as energy use) to both known and as yet unknown threats and challenges. Successful action often links local action with global networks, producing the cross-scale effects and multi-level change phenomena recognised by social-ecological systems theorists.

Reference

Berkes, F. & H. Ross (2013). Community resilience: Toward an integrated approach. *Society and Natural Resources* 26(1): 5-20.

3.1. Learning from Co-Founders of Grassroots Initiatives: Personal Resilience, Transition, and Behavioral Change – a Salutogenic Approach

GESA MASCHKOWSKI, NIKO SCHÄPKE, JANINA GRABS, NINA LANGEN

3.1.1. Introduction

The societal transformation toward a sustainable and low carbon society faces a number of technical and structural challenges. But it has become more and more obvious that among the main barriers blocking the desired change are basic 'mental infrastructures' - that is, how the promise of infinite economic growth has been embraced in the minds and hearts and in the hopes and dreams of Western societies.[66] Sustainability and climate change discourses reveal the inherent tension between this desire for permanent economic growth and the claim for a fair division of the Earth´s resources within and across generations.[67] People often find themselves in moral predicaments when ecologically harmful practices are invested with worthy purposes through social, national, and economic justifications.

Moreover, the situation is characterized by displacement and diffusion of responsibility.[68] Some attribute the responsibility to take action to governments whereas others attribute it to consumer citizens.[69] As a result, the vast majority of citizens do not engage sufficiently into pro-environmental behaviour.[70] On the contrary, environmental campaigning and media

66   Welzer, H., 2011. *Mental Infrastructures: How Growth Entered the World and Our Souls*. Berlin: Heinrich Böll Foundation.

67   Jackson, T,. 2009. *Wirtschaft ohne Wachstum*. Munich: Oekom.

68   Bandura, A. 2007. Impeding ecological sustainability through selective moral disengagement. *Int. Journal of Innovation and Sustainable Development* 2(1):8–35.

69   Grunwald, A. 2010. Wider die Privatisierung der Nachhaltigkeit – Warum ökologisch korrekter Konsum die Umwelt nicht retten kann. [Against Privatisation of Sustainability – Why Consuming Ecologically Correct Products Will Not Save the Environment] *GAIA - Ecological Perspectives for Science and Society* 19(3):178-182.

70   Osbaldiston, R. & J.P. Schott. 2012. Environmental Sustainability and Behavioral Science: Meta Analysis of Proenvironmental Behavior Experiments. *Environment and Behavior* 44(2):257-299.

coverage about causes and effects of climate change can provoke a backlash and 'climate fatigue', 'eco-anxiety or 'post-petroleum stress disorder'.[71] Kenis and Mathijs observed several obstacles for civic engagement such as the feeling of powerlessness, 'strategy scepticism' and resistance towards being 'conditioned' by awareness-raising campaigns.[72]

In this situation characterized by complexity and unclear responsibilities there exist rising numbers of people and groupings that search for alternative answers peripheral to what can be called the 'mainstream'. They take responsibility and experiment with sustainable ways of life: in food production (e. g., Consumer Supported Agriculture, Consumer Supported Enterprises), energy (Energy Cooperatives), transportation (Car sharing, free public transport) but also regarding new economic concepts and projects for a post-growth economy (e.g. Gift Economy, local currencies, 'REconomy' projects). According to Seyfang and Smith, these so-called grassroots movements are

> *"[I]nnovative networks of activists and organisations that lead bottom-up solutions for sustainable development; solutions that respond to the local situation and the interests and values of the communities involved. In contrast to the greening of mainstream busi-ness, grassroots initiatives tend to operate in civil society arenas and involve committed activists who experiment with social innovations as well as using greener technologies and techniques."[73]*

On the level of societal niches grassroots movements adopt the role of change agents. Recent research has demonstrated that change agents in general and engaged citizens in particular can initiate societal processes of change and contribute to the transformation of societies, provided that certain motivations, competencies, activities, learning processes and structural frame-conditions concur.[74]

Research on the success conditions of grassroots initiatives reveals that the number of parti-cipants is an important factor influencing the success of initiatives or the so-called ´upscaling´ of movements.[75] An interesting research direction therefore is to understand better the

71   Kerr R. A., 2009. Amid worrisome signs of warming, 'Climate Fatigue' sets in. *Science*, 326(5955): 926-928. Doherty T.J. & S. Clayton, 2011. The Psychological Impacts of Global Climate Change. *American Psycholo-gist* 66 (4), 265–276 Hopkins, R., 2008. *The Transition Handbook: From Oil Dependency to Local Resilience*. Totnes: Green Books.

72   Kenis A. & E. Mathijs, 2012. Beyond individual behaviour change: the role of power, knowledge and strategy in tackling climate change. *Environmental Education Research* 18(1):45–65.

73   Seyfang, G. & A. Smith, 2007. Grassroots innovations for sustainable development: towards a new research and policy agenda. *Environmental Politics* 16(4), p 585.

74   Ornetzeder M. & H. Rohracher, 2013. Of Solar Collectors, Wind Power, and Car Sharing: Comparing and Understanding Successful Cases of Grassroots Innovations. *Global Environmental Change*, 23(5):856-867. WBGU, 2011. World in Transition – A Social Contract for Sustainability. Flagship Report, German Advisory Council on Global Change (WBGU). Berlin: WBGU, Pp 241ff. Kristof, K. 2010: *Wege zum Wandel. Wie wir gesellschaftliche Veränderungen erfolgreich gestalten können*. Munich: Oekom, p 520.

75   Feola G. & Nunes. R.J. 2013. *Failure and Success of Transition Initiatives: a study of the international replication of the Transition Movement', Research Note* 4. Walker Institute for Climate System Research, University of Reading, August 2013. Seyfang, G. & A. Smith (eds.). 2013. Grassroots Innovations. *Global Environmental Change*. Special issue, Vol. 23. Middlemiss, L. & B. Parrish. 2009. Building capacity for low-carbon communities: The role of Grassroots initiatives. *Energy policy* 38: 7559-7566.

preconditions for engagement. Kristof points out that the essential psychological difference between change agents (those already active) and their target audience are their experiences and the associated progress they have already made.[76] They have passed through cognitive, motivational, and behavioral developments that made them change agents.

Perspicuous as this appears, it brings up a number of additional questions with regard to a deeper understanding of the processes through which people become change agents. This research is therefore motivated by the following questions:

i. Why do grassroots actors behave differently from the majority; what motivates them to engage and start an initiative?

ii. What can we learn from grassroots innovators with regard to causes and conditions of civic engagement?

iii. To what extent is it possible to 'mainstream' these determinants?

To develop a deeper understanding of the psychological processes through which people become change agents, we adopted a qualitative case study approach to analyze three different German grassroots movements. We assumed that their engagement can be interpreted as a healthy reaction, a form of (self-) empowerment confronted with a complex and frightening situation which causes widespread human harm and environmental degradation. To analyze their engagement we relied on the concept of salutogenesis. The concept of salutogenesis is related to positive psychology and personal resilience.

> *Grassroots engagement could be interpreted as a healthy reaction, a form of (self-) empowerment confronted with a complex and frightening situation which causes widespread human harm and environmental degradation.*

Subsequent sections are structured as follows. Section 2 presents an overview of the salutogenic concept. Section 3 summarizes our aims and research questions. Section 4 provides a brief summary of the three cases and our research method. Section 5 presents our initial findings. In section 6 we discuss some insights on the potential and the limitations of grassroots movements for social transformation. Section 7 concludes by highlighting the suitability of the salutogenetic approach for understanding grassroots engagement. Applying the

---

76   Kristof, K. 2010. *Wege zum Wandel. Wie wir gesellschaftliche Veränderungen erfolgreich gestalten können.* Munich: Oekom, p 515.

Figure 3.1.1 – *Potato Harvest 2014, CSA-Bonn. Credit: Gesa Maschkowski.*

approach points towards the need for rethinking the aims that founders of initiatives pursue and considering what are the most promising levers for upscaling grassroots movements.

3.1.2. Salutogenesis: Why?

The term and theory of salutogenesis were developed in the 1970s by the medical sociologist Aaron Antonovsky. The word salutogenesis consists of the Latin term 'salus' (health, well-being) and the Greek word 'genesis' meaning emergence or creation. In his work, Antonovsky discovered that some people stay healthy despite traumatic experiences such as imprisonment in a concentration camp or flight during wartime. This observation evoked a shift in his intellectual orientation from looking at risk factors of health to the identification of the strengths of an individual. A salutogenic orientation does not analyze why people get sick. Rather, it addresses the question, 'What explains the movement toward the health pole of the health ease/dis-ease continuum?'. According to Antonovsky, health is not a static condition; rather, it can be seen as a continuum that ranges from complete well-being to total dysfunction. The central factor which enables humans to overcome the omnipresent external and internal stressors and stimuli is the Sense of Coherence (SOC).

The SOC is a "way of looking at the world", defined as an enduring but flexible...

> "...feeling of confidence that (a) the stimuli deriving from one´s internal and external environments in the course of living are structured, predictable and explicable; (b) the resources are available to her/him to meet the demands posed by these stimuli; and (c) these demands are challenges worthy of investment and engagement."[77]

He thereby merges three different psychological factors: comprehensibility on a cognitive level; manageability on a behavioral level; and meaningfulness on an emotional level, in other words the sense that, "[L]ife is worth the effort, it's meaningful and creates happiness."[78] Figure 3.1.2 summarizes these interrelations.

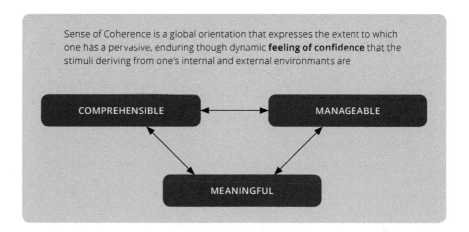

Figure 3.1.2 – *Dimensions or Sense of Coherence (Based on Antonovsky 1997).*

The sense of coherence (SOC) concept has been applied successfully in biomedicine and public health research to assess personal resilience in areas such as health promotion, education, working life, and war and post-conflict settings.[79] Bengel et al. have demonstrated that SOC is essentially a construct of psychological health and is linked to personal resilience.[80]

77   Antonovsky, A. 1997. Salutogenese. *Zur Entmystifizierung der Gesundheit*. DGVT, Tübingen.

78   Antonovsky A., 1996. The salutogenic model as a theory to guide health promotion. *Health Promo Intl*. 11(1):11-18.

79   Almedon, A.M., Tesfamichael, B., Saeed Mohammed, Z., Mascietaylor C. G. N. and A. Zemui, 2007. Use of ´sense of coherence (SOC)´ Scale to measure Resilience in Eritrea: Interrogating Both, the Data and the Scale. *Journal of Biosocial Science* 39 (1): 91-107. Eriksson M. and B.J. Lindström, 2007. Antonovsky's sense of coherence scale and its relation with quality of life: a systematic review. *Epidemiol Community Healt*. 61(11): 938–44.

80   Bengel J., Strittmatter R. and H. Willmann, 2001. *Was erhält Menschen gesund? Antonovskys Modell der Salutogenese - Diskussionsstand und Stellenwert*. Köln: BzgA.

Other results show positive relations between SOC and self-efficacy,[81] and between SOC and quality of life.[82] These findings have led to the integration of the salutogenic approach into competence-oriented health promotion and education.[83]

3.1.3. Goal and Research Questions

To the authors' knowledge, the theory of salutogenesis has not yet been applied to new social movements. In this paper, we use it to analyze why and how protagonists in grassroots movements manage to engage for change despite a situation marked by uncertainty and unclear responsibilities. We operationalize the concept of salutogenesis as the following core research questions:

› Comprehensibility: How do founders of grassroots initiatives understand and explain the current problems and challenges of our society?
› Meaningfulness: Why do they believe that their commitments make enough sense to be worth the effort?
› Manageability: Why do they feel capable of making a difference?
› Quality of Life: How does the commitment of the actors influence their perceived quality of life?

3.1.4. Cases Studied and Methods Used

This exploratory case study focuses on six key persons, co-founders of three grassroots movements present in Germany, namely:

i. *Carrot Mob Cologne*: organisers of temporary 'buycotts' in the form of purchase flash mobs by a crowd of carrot mobbers. They buy a lot of goods from one company in a small time period to encourage sustainable business behavior and initiate substantial carbon reductions in the selected (food) retail market;
ii. *Community Supported Agriculture* (CSA) in Bonn: a locally-based economic model of food production and distribution that directly connects farmers and consumers;
iii. The *Food Sharing and Food Saving Network*: tackling food waste by facilitating the non-monetary exchange of ´to-be-wasted´ foodstuffs between private persons and groceries, retailers and supermarkets;

A more in-depth explanation of the three initiatives can be found in Table 3.1.1.

---

81   Kröninger-Jungaberle, H. & D. Grevenstein, 2013. Development of salutogenetic factors in mental health - Antonovsky's sense of coherence and Bandura's self-efficacy related to Derogatis'symptom check list (SCL-90-R). *Health and Quality of Life Outcomes* 11(80):3-9.

82   Eriksson M.& B.J. Lindström, 2007. Antonovsky's sense of coherence scale and its relation with quality of life: a systematic review. *Epidemiol Community Healt.* 61(11): 938–44.

83   Krause, C., 2011. Der salutogenetische Blick. Fachstandard in der Arbeit von Erzieher/innen? In: Textor M.R.(ed.) *Kindergartenpädagogik- Online-Handbuch.* http://www.kindergartenpaedagogik.de/dg.html. Methfessel, B., 2007. Salutogenese – ein Modell fordert zum Umdenken heraus. *Ernährungs-Umschau* 54:704-709.

Table 3.1.1 – *Comparative Overview of the Three Movements.*

	CARROT MOB (CM) COLOGNE	FOODSHARING GERMANY AND FOODSHARING COLOGNE	CSA BONN
Founded in	2009	2012	2013
Level of activity	City level	City level and nationwide	City and its surroundings
Success measures	Share of turnover invested in CO_2 savings, 6 mobs in Cologne between 2010-2012, currently limited activity	Almost 35,000 kg food saved (03.06.2014)	Various: e.g. creation of 2.5 workplaces, high levels of satisfaction among members, organization of 7 community events per year, 6-7 voluntary work assignments per member per year
Core group characteristic	6 core members. In the four weeks before a mob: 50 hours per week	20 persons working 200 hours per week on voluntary basis	10 persons working 80 hours per week on voluntary basis
Number of members	Large number of 'mobbers' (up to 350). 350 followers on Twitter	3,500 activists saving food, 40,000 users registered online, 660 traders and manufacturers	120 members
Budget, annual	No budget. Flyers etc. financed by donations and the team	€40,000 for external services such as programming work, posters etc.	€105,000 for organic agriculture, including €2,000 for the expenses of the core group
Organisation form	No formal organization, voluntary work	Registered charity, voluntary work	No formal organization voluntary work

Hypotheses were generated on the basis of a broad literature review on grassroots movements' motivations and success factors and drawing on different theories. Based on our hypotheses we established and tested a semi-structured interview schedule, including questions about the main components of salutogenesis described above. Analysis of interview data took an inductive approach, using the principles of qualitative content analysis,[84] meaning that inductive codes were formulated step by step out of the material. We subsequently analysed connections between the codes and the components of the concept of salutogenesis using Atlas.ti 7.1.8 software for computer-based analysis of qualitative data.

84   Mayring, P., 2000. Qualitative Content Analysis. *Forum Qualitative Sozialforschung / Forum: Qualitative Social Research*, 1(2), Art. 20, *http://www.qualitative-research.net/index.php/fqs/article/ viewArticle/1089/2385*.

In the following section we present initial findings, organised into four categories that map onto the four components of salutogenesis. The category 'comprehensibility' includes statements that explain problems with our current societal system: their dimensions, individual evaluations, and feelings and thoughts on the topic. The category 'meaningfulness' is based on answers to the question, "Why is your engagement worth the effort?", along with relevant responses elsewhere in interviews. We define 'manageability' as the extent to which a person believes that he or she can mobilise the resources necessary to execute a project successfully. This category encompasses answers to the question, "Why do you feel capable of making a difference?". It also includes other factors enhancing self-efficacy mentioned by the interviewees, such as attitudes. Quality of life (QoL), the fourth category, includes statements about the effects engagement has on interviewees' perceived quality of life. QoL is not directly a part of SOC. Longitudinal studies confirm the predictive value of SOC for a good QoL: the stronger the SOC, the better the QoL.[85] We therefore used QoL as control variable.

3.1.5. Results

Comprehensibility: How do Founders of Grassroots Initiatives Explain the Current Challenges?
Interviewees highlighted two different problem areas. The first consists of problems that describe tangible negative effects of our present economic and societal systems on the environment and people, such as lack of resources, waste of food, inequality, population increase and globalization.

> *"I just said trash, CO_2, overpopulation, sealing of the soil, heaps of things have been going on - over indebtedness - it is LUNATIC a lot of what has been going on."*
> (CSA, P1:64)

In particular, co-founders of the German foodsaving movement stressed the systemic tendency of the current monetary system towards increasing injustice and obligating people to act against their values and needs. The massive slaughter of animals, for example, would never be done voluntarily. It happens because people are paid for doing it:

> *"If we somehow boil it down to the people who like to kill animals while aiming to do society a huge favor, there would be far fewer people doing it than at the moment."*
> (Foodsharing 2, P8:42)

The second problem area emphasised in the interviews is the way our society deals with the aforementioned problems, including system-based constraints and anxieties.

---

85   Eriksson M.& B.J. Lindström, 2007. Antonovsky's sense of coherence scale and its relation with quality of life: a systematic review. *Epidemiol. Community Healt.* 61(11): 938–44.

"It is like a huge clockwork [device] and everybody is caught in their own little cogwheel, everything is entangled and it is so darned difficult to get out, because the wheels are turning and turning and you yourself are only a small cog in a big machine. And you have to try very actively and consciously to stop your cogwheel and step out of the system. And when you have taken the first step, the next one is easier, but the majority of people is caught in the cogwheel ...and dependency and fear as well..."
(CSA, P3:44)

One aspect that many interviewees mentioned is widespread ignorance and even suppression of the problems, meaning people accept and make do with the current state of the system, instead of acknowledging the problem and taking responsibility for changing the situation. This leads to the acceptance of conditions that violate basic social values.

"... through Foodsharing many [managers of] grocery stores for the first time realized how much they actually throw away. The employees knew this, since they were doing it on a daily basis. But, since there were no numbers, no transcripts, this was sacrificed for a commercial logic: the shelves must always be filled ..."
(Foodsharing 1, P6:58)

Feelings in the Face of the Problems

Frustration, powerlessness, anxiety, monotony – these were feelings associated with the predominant problems. Anger was the most commonly-mentioned emotion. It arose, for instance, when the current system was perceived to be violating strongly held personal values of interviewees (e.g. on the nature of food):

"Good food and good drinks and food culture in itself has always fascinated me and the more I understood about the food industry over the years the more repulsed and also angry I got (...). That is our source of life. We are what we eat and if we don't produce our food in a sustainable way we destroy everything and we destroy the soil, we destroy our energy resources through influences of the market."
(CSA 2, P3:20)

Many interviewees perceived problems as complicated and overwhelming:

"It´s maddeningly complicated [...] you never understand everything. At least you have to develop a certain attitude, without saying, 'I have to become an expert' because you can only be an expert on a small area, you have to try to keep an overview."
(Foodsharing 1, P6:54)

Meaningfulness: Why do Co-founders of Grassroots Initiatives Consider these Challenges Worthy of Investment and Commitment?

From interviewees' responses we could infer strategies to cope with challenging problems such as the wish to act in accordance with intrinsic values and to stay authentic and credible.

> *"To keep face and to not go with the flow and well, to create also for oneself a livable life, a livable environment, and a livable structure, that is the change I want to achieve."*
> (CSA 2 P3:60)

> *"So I really do only what I am convinced of 100 percent."*
> (Foodsharing 2, P8:48)

Other major sources of meaningfulness were positive emotions, such as joy and fun, arising from connecting positive visions with positive action and building of social cohesion:

> *"In the CSA and in the Transition Town Movement you do something that creates joy. It is a positive impulse. It is not going against [the mainstream], rather just doing it differently, without asking the politicians, this is fascinating. That you can choose a different way in a society and just realize it. And that by these means, a lot of other things are made possible."*
> (CSA 2, P3: 25)

> *"The main reason [for my engagement] is, as I just explained, the personal contact, the people."*
> (CSA 1, P1: 33)

Personal engagement is assessed as valuable because it is a way to reach people in different social classes. Common sources of motivation mentioned include the wish to create aware-ness and to give a thought-provoking impulse in order to encourage rethinking. This aware-ness-impulse was in some cases classified as more important than the project itself.

> *"When the entrepreneur AFTER the activity said, 'Hey that was awesome', and his turnover was increased and he would now look forward to new investments also on matters of the environment. THAT has persuaded me more and more. (...) One could easily notice: something moved in his head."*
> (Carrotmob 1, P5:74)

> *"And I believe that something is happening in the background which is AT LEAST just as important, that everyone who shares their food or receives food gets their mind nudged about why we throw so much away. And they start to reconsider their consumer behavior. Essentially, that is an important goal for me because the mere distribution of excess food is not a proper solution in the strict ecological sense."*
> (Foodsharing 1 P6:58)

Manageability: What Makes People Confident they have Access to the Resources Necessary to Meet the Challenges?

Interviewees mentioned numerous factors that support them to feel capable of making a difference. These include:

› External factors such as positive role models, best practice examples, supporters and mentors:
  "There are enough positive examples, why shouldn´t we make it?" (Laughs out)
  (CSA 2, P3:64)
› Personal factors such as previous positive life experiences with change (mastery experiences)
› Strategies such as expectation management, meaning that anticipated and targeted results are concrete, feasible and realistic:
  "I am soberingly realistic… I do not expect that my action has so much impact, that's a great relief."
  (CSA 2, P3: 59)
  "I start with small baby steps and I don´t have the feeling of being able to change much […], but it's fun to work together with the group, to work together with the farmer and it would be even more fun to work on the field."
  (CSA 1, P1:77)
› Positive experiences and emotions associated with the engagement itself, in particular positive group processes and complimentary feedback:
  "This is the first time that I feel a great success by reaching people, but also for myself, that I am happier instead of getting annoyed with something."
  (Foodsharing 2, P8:72)

When we asked about conditions necessary for up-scaling movements, interviewees stressed the importance of enabling other people to gain positive and concrete experiences, for instance by providing low-threshold opportunities to engage:

  "People have to be invited; I think […] therefore, the Foodsharing and Foodsaving movement is a good starting point. It is practical, it is easy to understand, and when people deal with that problem they can recognize that this concrete example is only a symptom. And then, they can look for the causes of the problem."
  (Foodsharing 2, P8:392)

Attitudes for Action

Interviewees mentioned several attitudes associated with personal engagement, namely: Non-conformity or radicalism, naivety, curiosity, cooperation instead of domination, healthy confidence, courage and a healthy megalomania, and feelings of responsibility.

Quality of Life

Our examination of the SOC would be incomplete if we omitted the question of how dedication affects the quality of actors' lives. Answers were comparatively simple. When asking the question, "Do you have the feeling that you are giving up something due to your commitment?"

all interviewees reacted with surprise and disagreement, and highlighted the positive sides of their dedication such as:

› Creativity and learning
› Feelings of connectedness to the city and the people
› New social relations
› Pleasure and health following the motto 'less is more'

Restrictions of the freedom of choice, such as in food, was either ranked as insignificant or even as time saving and a relief. It is also notable how often positive feelings were mentioned when interviewees explained personal experiences they had during their engagement.

> *"I think it is totally beautiful and I am totally happy that I am able to be part of this.*
> *Having taken this step and being able to initiate this. Well, being able to bring the topic here."*
> (Foodsharing 2, P8:253)

Interviewees mentioned positive feelings such as luck or enthusiasm far more frequently than they did feelings such as anger or frustration. The latter did come up when interviewees talked about root problems. This points towards a key result of the engagement: engaging in the grassroots initiatives in a manageable, meaningful and thus salutogenic way seems to be correlated with a higher quality of life.

3.1.6. Discussion

The characteristics of the grassroots movements analyzed are different. Dimensions of variation include their target groups (customers, retailers, or citizens), their budget (none to several thousand Euros per year) and their scale (city-wide to nation-wide). Analysis of interviews, however, reveals a number of common patterns.

Comprehensibility and the Limits of Cognitive Knowledge

The collective action frame of the grassroot activists described the conflict between the growth paradigm on the one side and the exhaustion of earthly resources on the other. Even though interviewees did not explicitly identify themselves as members of global justice movements, the issues mentioned showed a close connection to these.[86] Co-founders of the initiatives felt anxious and powerless considering the very large dimensions of current social and ecological problems. Similar results have already been reported in the literature.[87]

Despite these feelings, grassroots actors were still able to engage for change. This leads to our research question of, "How do they achieve comprehensibility, as the ability to make

---

86   Schlichting, I. & A. Schmidt. 2012. Strategische Deutungen des Klimawandels. Frames und ihre Sponsoren. *Forschungsjournal Soziale Bewegungen* 25 (2): 29–41.

87   Kenis, A. & E. Mathijs, 2012. Beyond individual behaviour change: the role of power, knowledge and strategy in tackling climate change. *Environmental Education Research* 18(1):45–65.

sense of extreme and stressful events?" It is notable that the co-founders of grassroots initiatives relied on attitudes and values to explain and structure the problems, as mentioned by Foodsharing 1, "...you [will] never understand everything. At least you have to develop a certain attitude." From a salutogenic point of view the positive deviance of grassroots actors, i.e. taking action in the face of widespread ignorance and/or apathy, seems to rely on values and attitudes rather than purely cognitive assessments of problems. This finding is supported by Kay Milton´s work who argues that, "[T]he emotional and constitutive role of nature and natural things has been underplayed in western environmental debates, which have been dominated by a rationalist scientific discourse in which emotion is suppressed and emotionalism denigrated."[88] Our findings underscore the crucial role of emotions as the link between the appraisal of a situation and the motivation to take action (see Figure 3.1.3).[89]

Meaningfulness and Quality of Life Through Civic Engagement

The aims of the three initiatives as expressed by their co-founders challenge the ideological foundation of the consumer society which still adheres to the narrative of, *'The more we consume the better off we are.'* In the latter line of thought, demands to reduce the material impact of human activities are likely to be perceived as constraining human welfare and threatening quality of life.[90] Accordingly, co-founders of grassroots initiatives can be expected to suffer from lower quality of life, due to their reduced consumption and time-consuming activism.

The results of the study however show that different narratives and effects prevail: interviewees reported that it is gratifying to act in ways consistent with their own values. Personal commitment and assumption of responsibility for one's own environment leads to empowerment and social learning for oneself and others. Social capital is created by social cohesion in initiatives and connections among people who would not otherwise have met. Interviewees were inspired by the possibility to create awareness and initiate small system changes. 'Meaningfulness' in the salutogenic sense was created by their attempts to express positive and constructive attitudes and values in their own lives. Tensions between the inability of the current economic system to reproduce fundamental values such as justice and human rights and the desire of

> *'Meaningfulness' in the salutogenic sense was created by their attempts to express positive and constructive attitudes and values in their own lives.*

88 Milton, K. 2002. *Loving Nature: Towards an Ecology of Emotion*. London: Routledge, p.91

89 Klandermans, B. 2004. The demand and supply of participation: Social psychological correlates of participation in a social movement. Pp. 360-379 in: Snow D.A., Soule, S. & HP Kriesi (eds.) *Blackwell Companion to Social Movements*. Oxford: Blackwell.

90 Jackson, T., 2005. Live better by consuming less? Is there a double dividend in sustainable consumption? *Journal of Industrial Ecology* 9(1–2): 19–36.

Personal Drivers of Engagement

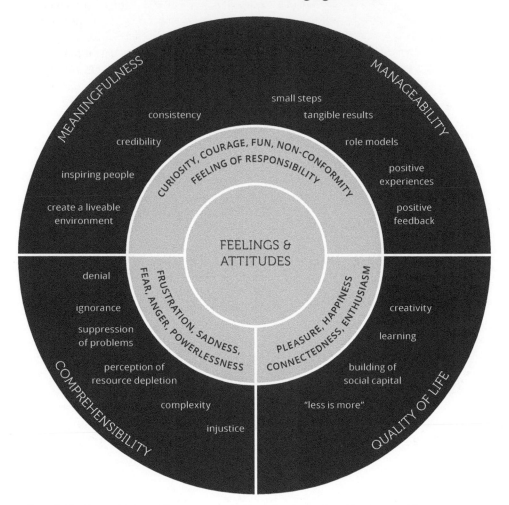

Figure 3.1.3 – Personal Drivers of Engagement in the Dimensions Comprehensibility, Manageability, Meaningfulness and Quality of Life.

grassroot actors to act according to these values were resolved by their refusal to tolerate the situation and the assumption of personal responsibility for change. In that regard, founders acted in the tradition of ´classic´ social movements.[91]

By cultivating inner consistency, building social capital and initiating processes of social learning, the grassroots actors reported to have improved their perceived quality of life.

91   Della Porta, D. & M. Diani, 2006. *Social movements: an introduction*. Malden, MA: Blackwell. Pp. 64ff.

Recent research similarly showed that commitment to a Transition Together Initiative had positive effects on various aspects of health and well-being,[92] primarily attributed to community engagement and collaborating with immediate neighbors. This relates to Tim Jackson's suggested 'double dividend' potentially inherent in sustainable consumption: the ability to live better by consuming less, to reduce our impact on the environment and to become more human as a result. Nevertheless, Jackson himself is cautious about the prospects for a double dividend and draws attentions to the role of societal frameworks: "Such 'win-win' solutions may exist but will require a concerted societal effort to realize [on a broader scale]."[93]

Manageability: Enhancing Self-Efficacy through Concrete and Collective Action
Conditions supporting manageability include factors around behavior change highlighted as relevant also by other theories relevant in the field of behaviour change such as the Social Cognitive Theory developed by Albert Bandura[94]. Our research findings demonstrate the high relevance for mastery experiences, regarded in the SOC approach as the strongest influence on perceived self-efficacy.[95] Given the complex and frightening social and ecological problems that are mentioned in 5.1, realistic expectations regarding potential outcomes also seem to be very important. Feasible action plans allow for the transformation of frustration and anxiety into motivation and a sense of achievement. This finding is in accordance with research of Kristof, who reported that establishing concrete and realistic steps is a necessary prerequisite for reducing fear of transformation and change.[96]

Interviewees also mentioned best practice examples and role models as supporting factors. This so called ´social modeling´ is reported to be an important strategy for behavior change, successfully used in health and environmental education[97]. Grassroots networks, such as the Transition network, the Network of Consumer Supported Agriculture, Foodsharing Network or the Carrotmob Network fulfil an important role by providing these examples and models.

Finally, our research shows that grassroots initiatives provide opportunities to build positive emotions, for instance through group processes and collective action. Improving the emotional state is also regarded as an important strategy for increasing beliefs of self-efficacy.[98]

92   Richardson, J., Nichols, A. & T. Henry, 2012. Do transition towns have the potential to promote health and well-being? A health impact assessment of a transition town initiative. *Public Health* 126(11): 982-9.

93   Jackson, T., 2005. Live better by consuming less? Is there a double dividend in sustainable consumption? *Journal of Industrial Ecology* 9(1–2): 19–36.

94   Bandura, A., 2002. Environmental sustainability by sociocognitive deceleration of population growth. Pp. 209-238 iøn Schmuch P. & W. Schultz (eds.). *The Psychology of Sustainable Development*. Dordrecht, The Netherlands: Kluwer.

95   McAlister, A., Perry, C. & G. Parcel, 2008. How individuals, environments and health behaviors interact. In K. Glanz, B. Rimer, and K. Viswinath (eds.) *Health Behavior and Health Education*. Jossey-Bass, San Francisco, Calif, USA, 4ᵗʰ edition: 167-188.

96   Kristof, K. 2010. *Wege zum Wandel. Wie wir gesellschaftliche Veränderungen erfolgreich gestalten können*. Munich: Oekom, p 542.

97   see footnote 31.

98   McAlister, A., Perry, C. & G. Parcel, 2008. How individuals, environments and health behaviors interact. In K. Glanz, B. Rimer, and K. Viswinath (eds.) *Health Behavior and Health Education*. Jossey-Bass, San Francisco, Calif, USA, 4ᵗʰ edition: 167-188.

Moreover, positive action provoked positive feedback loops, enhancing feelings of manageability and contributing to empowerment of those involved.

The study however should not create the impression that constraints, failures, and negative experiences are absent from grassroots initiatives.[99] A common example is when scarcity of resources such as time and money provokes the decomposition of an initiative. In the case of the Foodsharing initiative it led to a sustained effort to be independent from money in order to achieve more autonomy and free up time for grassroots work. Group dynamics are another crucial factor, not

> *Positive action provoked positive feedback loops, enhancing feelings of manageability and contributing to empowerment of those involved.*

only with regard to success as already mentioned, but also in relation to failure of initiatives. This requires further analysis, beyond the scope of this paper.

3.1.7. Conclusion: Potentials and Limitations of Grassroots Initiatives

In this study we aimed to understand reasons for the engagement of individuals in grassroots initiatives. We understood this engagement as a ´salutogenetic´ process: a healthy reaction to being confronted with the intellectually and morally overwhelming situation of current unsustainability. To understand the psychological processes underlying this positive action, and thus identify potential levers for upscaling engagement in initiatives, we used Antonovsky's concept of salutogenesis. According to Antonovsky, the ability of a person confronted with a major challenge to take constructive action depends on the sense of coherence they are able to maintain. This sense of coherence in turn can be broken down into aspects of manageability, comprehensibility and meaningfulness. We used these concepts to analyze interviews with founders of grassroots movements.

A core result of this study is that the use of the concept of salutogenesis can provide a deeper understanding of psychological factors motivating change agents to initiate grassroots movements. Founders of initiatives perceive the given, unsustainable situation as a challenging, potentially frightening one. They try to stay healthy and active in this situation by developing meaningful engagement, based on a comprehensive interpretation of the given situation and challenges, and taking manageable actions. In particular, the salutogenetic approach as applied here may contribute the following three insights to the discourse:

99 Feola, G. & R.J. Nunes, 2013. *Failure and Success of Transition Initiatives: a study of the international replication of the Transition Movement', Research Note* 4. Walker Institute for Climate System Research, University of Reading, August 2013.

1. The scale and complexity of current problems are beyond the scope of attempts at purely cognitive explanation. The ability to comprehend associated challenges in a salutogenic way is connected to the role of attitudes and values, helping actors to explain and organize otherwise overwhelming information. Learning from grassroots co-founders as persons acting in a salutogenic way would imply that people need opportunities to (re-) connect with internal values and to act accordingly. At this point, one should consider to what extent expert-dominated discourses about climate and social change, and associated striving for objectivity, are both patronizing and constitute a potential obstacle to ´*comprehensibility*´ in itself: are they hindering the (re-) connection with attitudes and values necessary to understand, explain and deal with the problems? Do we need a shift of awareness from detached observers to engaged participants, from objectivity to critical subjectivity?[100] And would this critical subjectivity be needed in many parts of society - from citizens, to politics, the media and science?

2. The *manageability* of action on current challenges is enhanced by positive action, realistic aims, and best practice models, and in particular by positive group processes. This is a completely different approach from current governmental strategies to foster sustainable lifestyles by addressing individuals in their roles as consumers. It raises the question of when, how and where the majority of people have the opportunity to make positive, collective experiences of change, enhancing self-efficacy and therefore feelings of manageability.

3. Interviews with co-founders of grassroots movements contain a great diversity of positive narratives explaining why it is joyful and *meaningful* to work for change. Those who hoped that wider society can learn from grassroots initiatives how to make consumerism 'greener' within our hegemonic social system will be disappointed on this point. The goal of the co-founders is not in the first place to change consumer behavior within the given system. They identify the system itself, its environmental and social problems, as targets of collective action. They create awareness of the need of systemic change and start to build alternatives on a niche basis. In this case, the change of consumer behavior is a positive and certainly desirable consequence but not the main motivation, which is the prospect of deeper societal change.

These insights have consequences concerning potential levers to upscale involvement in grassroots initiatives as a mechanism for societal change. To identify these levers, the very processes that allow initiatives to contribute to societal change need to be reconsidered: upscaling and mainstreaming the activities and experiences of the co-founders of grassroots initiatives would thus mean upscaling and mainstreaming opportunities for citizens to engage collectively, to shape their environments, and thus gain positive experiences by making small realistic steps with the support of others. Or in Otto Scharmer's words, "[By] *creat[ing]*

---

100   Sterling, S. 2007. Riding the Storm: towards a connective cultural consciousness. Pp, 63-82 in: Wals, A.E.J,
        2007. (eds.) *Social Learning Towards a Sustainable World, Principle, perspectives and praxis*. Wageningen:
        Wageningen Academic Publishers.

infrastructure innovations that allow all citizens to become aware of their real power in co-creating the intentional ecosystem economy and in deepening our democracy."[101] We are not, therefore, discussing the upscaling of green consumerism and some general green or sustainability engagement schemes applicable to defined contexts, but the upscaling of social learning and empowerment allowing people to take action in their particular way: manageable, meaningful and comprehensive. We therewith agree with recent discussions pointing towards the need of

Figure 3.1.3 – *First Carrotmob, Cologne. Credit: Martin Terber.*

101   Scharmer, O., 2009. Seven Acupuncture Points for Shifting Capitalism to Create a Regenerative Ecosystem Economy. Paper prepared for presentation at the: Roundtable on Transforming Capitalism to Create a Regenerative Economy MIT, June 8–9; Sept. 21, 2009. P.2. *http://www.ottoscharmer.com/sites/default/files/2009_SevenAcupuncturePoints.pdf*

scaling-up processes instead of objects or product designs[102] and to understanding societal transformation as a process of social learning.[103]

It would be an excessive demand, however, to place the responsibility for social transformation solely on the shoulders of citizens. Successful projects for lifestyle change have always been supported by multi-level approaches. This was impressively shown by the Finnish project 'Health in all Policies' which changed the food habits of the entire Finnish society over a 20 year time frame and led to tremendous reductions in mortality rates from cardiovascular diseases. The project pursued a community-based approach, accompanied by effective measures in public health care, social modeling, media campaigns (local and national), the economy (new products), and politics (taxes).[104] When aiming for widespread societal change it therefore could be indicative to ask the salutogenic question on every level: What do politicians, businesses, scientists, teachers, students, the administration, and citizens, or in other words, what does our society need to gain a deeper and simultaneously more flexible feeling of trust that the transformation towards a sustainable low carbon society is comprehensible, manageable and makes sense?

> *Upscaling and mainstreaming the activities and experiences of the co-founders of grassroots initiatives would thus mean upscaling and mainstreaming opportunities for citizens to engage collectively, to shape their environments, and thus gain positive experiences by making small realistic steps with the support of others.*

Acknowledgements

The research leading to this working paper was done as part of the research project "Close up on Grassroots", generously funded by the Ministry of Innovation, Science and Research of North Rhine-Westphalia and the Competence Centre for Consumer Research of North Rhine-Westphalia.

102 Smith A., 2014. *Scaling-up Inclusive Innovation: Asking the Right Questions?* http://steps-centre.org/2014/blog/scaling-up-inclusive-innovation/.

103 e.g. InContext http://www.incontext-fp7.eu/.

104 Puska P. & T. Ståhl, 2010. Health in All Policies — The Finnish Initiative: Background, Principles, and Current Issues. *Annual Review of Public Health* (31):315-28.

This article is based on the presentation "Personal Resilience, Transition and Behaviour Change - a Salutogenetic Approach" given in the Session "Resilience, Community Action and Social Transformation", 6th May 2014. Montpellier, France. This session was organized by the Transition Research Network and ECOLISE as part of the Resilience 2014 conference.

We would like to express our special thanks to Tom Henfrey for careful editing of the manuscript and his insightful comments and suggestions.

3.2. Resilience and Community Action in the Transition Movement

TOM HENFREY AND NARESH GIANGRANDE

3.2.1. Background, Definitions and Characterisations

Since its foundation, the Transition movement has held resilience as a key operational concept and stated goal. However, the actual meaning of 'resilience', and the implications of this for practice, is neither fixed nor consistent. No-one has attempted to put forward a single guiding definition, and its meaning and interpretation have changed over time as the movement matures, becomes established in new places, and encounters changing global conditions. Within Transition, resilience therefore means different things, to different people, in different places, at different times.

This paper initially emerged from attempts to identify suitable monitoring and evaluation methods and strategies for Transition groups and projects, and for the movement as a whole. The initial aim was to do this in relation to the stated goal of building resilience. It quickly became apparent that this was an unrealistic goal, at least in the short term and in relation to immediate practical needs.[105] As a concept, resilience is employed in a fluid and dynamic way, and so defies ready generalisation. As a goal, it is emergent over the long term in complex systems whose dynamics are markedly non-linear. This means that the short term changes that come about through resilience-building efforts do not necessarily reliably map onto the longer term changes that come about when the system shifts to a more resilient state.

Alongside this realisation, we became aware of a need to explore the relevance of resilience to community action more fully and critically. This piece of work presents some preliminary findings from that work. It examines relationships between approaches to and experiences of resilience within the Transition movement and their relationships with key theoretical

---

105 This line of work eventually morphed into the Monitoring and Evaluation for Sustainable Communities project, a partnership among Transition Network, the Low Carbon Communities Network, and the School of Geography at Oxford University: *https://mescproject.wordpress.com/*

treatments. In conclusion, it responds to attempts to appropriate the concept of resilience in the service of neoliberal agendas. It argues that this opens up a discursive space within which a scientifically and ethically grounded concept of resilience can act as a 'Trojan horse' by revealing inherent contradictions in the political economy of neoliberalism, and growth-oriented economics more generally.

> *The short term changes that come about through resilience-building efforts do not necessarily reliably map onto the longer term changes that come about when the system shifts to a more resilient state.*

It was a basic founding observation of the Transition Movement that meaningful responses to peak oil and climate change will, of necessity, transform a society whose economic stability relies on ever-increasing inputs of energy derived from inherently limited sources and accumulation of pollutants beyond the biosphere's capacity to absorb them.[106] Transition's practical programme is based on the conjecture that the nature and outcome of this transformation will depend on the extent to which communities take responsibility for it through pre-emptive action, and the recognition that it is also an opportunity to build resilience against future crises.[107]

Assessing Transition's characteristic early discourses on resilience against technical literatures, Haxeltine and Seyfang have identified three main weaknesses.[108] Firstly, there is an emphasis on resilience to specific threats – notably peak oil – and a lack of explicit recognition that responses to this may not promote resilience to other types of change.[109] Second, there is a tendency to equate resilience rather uncritically with localisation; extreme localisation may in fact reduce resilience in certain respects.[110] Third, there are costs associated with resilience-building – in particular, there seems to be a trade-off with the efficiency that has dominated economic policy in recent decades – so the ideal goal may be not to maximise resilience, but to achieve some necessary or optimum level.

106 Heinberg, R., 2004. *Powerdown: Options and Actions for a Post-Carbon World*. Gabriola Island, BC: New Society Publications. Jackson, T., 2009. *Prosperity without Growth*. London: Earthscan.

107 Hopkins, R., 2008. *The Transition Handbook: From Oil Dependency to Local Resilience*. Totnes: Green Books. Holmgren, D., 2009. *Future Scenarios: how communities can adapt to peak oil and climate change*. Totnes: Green Books.

108 Haxeltine, A., and G. Seyfang, 2009. *Transitions for the People: theory and practice of 'Transition' and 'Resilience' in the UK's Transition movement*. Tyndall Centre for Climate Change Research Working Paper 134.

109 Also see the distinction between general and specified resilience in Folke, C., S. R. Carpenter, B. Walker, M. Scheffer, T. Chapin & J. Rockström. 2010. Resilience thinking: integrating resilience, adaptability and transformability. *Ecology and Society* 15(4): 20. [online] URL: *http://www.ecologyandsociety.org/vol15/iss4/art20/*

110 North, P., 2010. Eco-localisation as a progressive response to peak oil and climate change – a sympathetic critique. *Geoforum* 41 (4): 585-594.

***Figure 3.2.1** – Cattle at Kattendorfer Farm, a CSA in Germany. Credit: Gesa Maschkowski.*

Subsequent developments have addressed these issues to some extent. The most thorough theoretical exploration of resilience directly in relation to Transition remains that in a doctoral thesis by Rob Hopkins, based on action research within Transition Town Totnes[111] undertaken in parallel with preparation of the Totnes Energy Descent Action Plan (EDAP).[112] This informs the ideas of resilience expressed in Hopkins' non-academic work on Transition,[113] which in turn are major influences on resilience thinking throughout the Transition Movement. It was also the basis for a set of 'Resilience Indicators' employed in the Totnes EDAP, which build upon resilience evaluation tools developed elsewhere. Many of these resilience indicators could equally be seen as localisation indicators; this to some extent upholds Haxeltine and Seyfang's second point above. The thesis also begins to move beyond localisation in consider the appropriate scale for various kinds of economic activity, arguing that some of these are best located at higher geographical scales.

111 Hopkins, R., 2010. *Localisation and resilience at the local level: the case of Transition Town Totnes (Devon, UK)*. PhD thesis, Plymouth University.

112 Hodgson, J. and R. Hopkins, 2010. *Transition in Action, Totnes 2030, an Energy Descent Action Plan*. Transition Town Totnes. *http://totnesedap.org.uk*

113 Hopkins, R., 2011. *The Transition Companion*. Green Books, Totnes, Devon. Hodgson, J. and R. Hopkins, 2010. *Transition in Action, Totnes 2030, an Energy Descent Action Plan*. Transition Town Totnes. *http://totnesedap.org.uk*

Hopkins identifies four major concepts of resilience in academic literatures relevant to theory and practice in Transition: [114]

› That of social-ecological resilience developed by the Resilience Alliance, as a system's capacity to maintain structural and functional continuity in the face of ongoing change.
› Dominant in the risk management literature, and prevalent in much government and related discourse on climate change adaptation, that of resilience as the ability to 'bounce back' after major shocks and/or crises, such as a natural disaster or terrorist attack.
› The notion of personal resilience, or a person's ability to cope with personal setback, hardship, trauma or other crises, commonly applied in social work, counselling and psychotherapy.
› The idea of community resilience, at the time developed largely as practical/evaluation tools by a range of organisations, more recently entering the formal scientific literature on resilience.

Sections 3.2.2 to 3.2.5 examine each of these in turn.

3.2.2. Social-Ecological Resilience

A typical definition of social-ecological resilience in the work of the Resilience Alliance is, "[T]he capacity of a system to absorb disturbance and reorganise while undergoing change so as to still retain essentially the same function, structure and feedbacks".[115] Originally based on observations of the self-maintaining properties of ecosystems,[116] this was later linked with insights from human ecology and ecological anthropology about how human societies negotiate ongoing change and inherent unpredictability in the ecological systems that provide their productive base.[117] Walker *et al* link resilience to two related capacities: adaptability, the capacity of human actors in the system to manage for resilience, and transformability, the system's capacity to undergo a fundamental reorganisation when changing circumstances mean it can not persist in its existing form.[118]

114   Hopkins, R., 2010. *Localisation and resilience at the local level: the case of Transition Town Totnes (Devon, UK)*. PhD thesis, Plymouth University.

115   Walker, B., C. S. Holling, S. R. Carpenter, & A. Kinzig, 2004. Resilience, adaptability and transformability in social–ecological systems. *Ecology and Society* 9(2): 5, p. 32. [online] URL: *http://www.ecologyandsociety. org/vol9/iss2/art5/*

116   Holling, Crawford S., L.H. Gunderson and G.D. Peterson, 2002a. Sustainability and panarchies. Pp. 63-102 in Gunderson and Holling (eds.). Holling, Crawford S., 1992. Cross-scale morphology, geometry, and dynamics of ecological systems. *Ecological Monographs* 62(4): 447-502.

117   Berkes, F. and C. Folke (eds.), 1998. *Linking social and ecological systems. Management practices and social mechanisms for building resilience*. Cambridge University Press.

118   Walker, B., C. S. Holling, S. R. Carpenter, & A. Kinzig, 2004. Resilience, adaptability and transformability in social–ecological systems. *Ecology and Society* 9(2): 5. [online] UøRL: *http://www.ecologyandsociety.org/ vol9/iss2/art5/*

Subsequent work revealed common patterns affecting resilience in social and economic as well as ecological and social-ecological systems.[119] It would nonetheless be naïve to assume that ecological resilience is identical to social resilience, or that either implies the other.[120] In particular, people provide capacities absent in non-human systems, including anticipating and planning for crises, modifying ecological properties and potential responses to these through technology, and enhanced abilities for learning and management.[121] Extending the scope to socio-technical systems, in which technologies (especially for energy conversion and use) have a major influence on flows of matter, energy and information, reveals further complexity.[122] Adaptability in social-technical systems depends on flexibility in both their technical aspects and the sets of formal and informal rules, agreements and customs that regulate their use.[123] Recharacterisation of this body of theory as evolutionary resilience maintains the emphasis on adaptive change identified in ecological studies while broadening the perspective to include insights from the social sciences, avoiding the risk that confusion over definition makes the concept of resilience effectively meaningless in the way some consider sustainability to have become.[124] This confusion in large measure derives from a failure to ground both academic and vernacular discourses on resilience in a sound understanding of resilience theory.

> *Walker* et al *link resilience to two related capacities: adaptability, the capacity of human actors in the system to manage for resilience, and transformability, the system's capacity to undergo a fundamental reorganisation when changing circumstances mean it can not persist in its existing form.*

119 Gunderson, L.H. and Crawford S. Holling (eds.), 2002. *Panarchy*. Washington DC: Island Press. Berkes, F., J. Colding and C. Folke (eds.), 2003. *Navigating Social-Ecological Systems: building resilience for complexity and change.* Cambridge University Press.

120 Adger, W.N., 2000. *Social and Ecological Resilience: are they related? Progress in Human Geography* 24(3): 347-364.

121 Gunderson, L., 2009. *Comparing Ecological and Human Community Resilience.* Community and Regional Resilience Initiative Research Report 5.

122 Smith, A., and A. Stirling. 2010. The politics of social-ecological resilience and sustainable socio-technical transitions. *Ecology and Society* 15(1): 11. [online] URL: *http://www.ecologyandsociety.org/vol15/iss1/art11/*

123 Genus, A., 1992. Social control of large-scale technological projects: inflexibility, non-incrementality and British North Sea oil. Technology Analysis & Strategic Management 4(2): 133-148. Genus, A., 2012 (forthcoming). The governance of technological transitions: the case of renewable energy. In G. Marletto (ed.), *Creating a Sustainable Economy*. London: Routledge.

124 Davoudi, S., 2012. Resilience: A Bridging Concept or a Dead End? *Planning Theory & Practice* 13(2): 299-307.

Holling, whose work on the structure and dynamics of ecological systems is the original basis of resilience theory, describes a four-stage *Adaptive Cycle* common to all ecological systems, which go through successive phases of growth, stability, decay and reorganisation/renewal.[125] In terms of resilience, the key phase is that of reorganisation, at which the system is particularly sensitive to the influence of both internal and external factors. Small changes in key variables during this phase may affect whether the system recovers its previous condition, shifts to a new and more desirable state, or breaks down into some simplified or degenerate form. Reorganisation is thus a time of crisis at which the system may change in unpredictable ways, and at the same time an opportunity for learning, where it may build in new features that strengthen its ability to respond to environmental change. The outcome of reorganisation – and hence the trajectory for subsequent renewal – often depends on context: whether and how the wider environment supports particular pathways of renewal.[126]

The outcome of reorganisation – and hence the trajectory for subsequent renewal – often depends on context: whether and how the wider environment supports particular pathways of renewal.

Holling's work is also the original source of important findings about resilience and scale. Complex systems are multi-leveled, incorporating nested adaptive cycles at different scales of space and time. Some cycles involve very localised, fast-changing variables; others operate over broader spatial areas and longer timescales. In a deciduous woodland, for example, individual leaves grow, live, drop and decay over timescales of months. Trees have lifetimes of decades, and parallel cycles in the lifetimes of woodlands or forest patches may operate over centuries. In any systems, these different scales have very different properties and must be described and analysed in different ways. Resilience is to a large extent an outcome of *Panarchy*, or the ways in which cycles at these different scales interact.[127]

Resilience theorists have identified two main types of cross-scale interaction. The *Remember* effect is where large and slow variables buffer the influence of smaller, quicker factors: for example, when a tree dies in a forest, the presence of other trees around it and in the seed-bank ensures another tree grows in its place. Such effects may also inhibit necessary change,

125 Holling, Crawford S., 1992. Cross-scale morphology, geometry, and dynamics of ecological systems. *Ecological Monographs* 62(4): 447-502. Holling, Crawford S. and L.H. Gunderson, 2002. Resilience and adaptive cycles. Pp. 25-62 in Gunderson and Holling (eds.). *Panarchy*. Washington DC: Island Press.

126 Folke, C., J. Colding and F. Berkes, 2003. Synthesis: building resilience and adaptive capacity in social-ecological systems. Pp. 352-387 in Berkes, F., J. Colding and C. Folke (eds.), 2003. *'Navigating social and ecological systems. Building resilience for complexity and change'*. Cambridge University Press.

127 Holling, Crawford S. and L.H. Gunderson, 2002. Resilience and adaptive cycles. Pp. 25-62 in Gunderson and Holling (eds.) *Panarchy*. Washington DC: Island Press.

Figure 3.2.2 – *Bicycle Lanes, Copenhagen. Credit: Gesa Maschkowski.*

as we are currently experiencing in relation to carbon lock-in, the interlinked linked social, technical and political/institutional barriers to a shift away from fossil fuel dependency.[128] The *Revolt* effect is when changes at small, fast scales escalate to higher levels while the latter are in their release phase, and thus become the dominant influence on the trajectory of the next regeneration phase.

Transformability often depends on the interplay between remember and revolt effects. The 2008 banking crisis originated when what were initially isolated actions of individual traders became pervasive in the financial system as a whole at a time when it was moving into a release phase.[129] Subsequent failure to reform global finance reflects a pernicious form of the Remember effect, akin to carbon lock-in, where established institutions, customs and norms conspire with self-interest among powerful actors to limit the possibility of desirable change. Transition and other social movements, in contrast, are seeking to build transformability by themselves becoming the seeds of positive change; the sources of future positive Revolt

---

128   Unruh, G., 2000. Understanding carbon lock-in. *Energy Policy* 28(12): 817-830. Mitchell, C., 2008. *The Political Economy of Sustainable Energy*. Basingstoke: Palgrave MacMillan.

129   Mellor, M., 2010. *The Future of Money*. London: Pluto Press.

Transition and other social movements, in contrast, are seeking to build transformability by themselves becoming the seeds of positive change; the sources of future positive revolt responses to crises at higher levels.

responses to crises at higher levels.[130] Adaptability in social-ecological systems inhabited by traditional resource users depends on management practices that mean different local areas are at different stages in the adaptive cycle at the same time, maintaining patchiness and heterogeneity at the level of the landscape.[131] In a similar way, innovation at small scales for sustainability in industrialised economies is the most likely source of future transformability, when the inherent unsustainability of economies whose stability relies on continuous financial growth make this a necessity.[132]

The vernacular notion that Transition, through numerous local efforts at building resilience, may have the emergent effect of promoting wider transformability – or of maximising the chances that inevitable transformations are for the better – is thus well grounded in resilience theory. This, however, assumes suitable conditions at higher levels. As the next section describes, predominant understandings of resilience at these levels reflect very different theoretical perspectives and ideological narratives.

3.2.3. Resilience and Disaster Response

Treatments of resilience in literatures on disaster response were until recently unconnected with theories of social-ecological resilience, and historically tended to emphasise two contrasting properties. First, what has been termed 'engineering resilience': the ability to resist change and hence maintain a constant state. Second, the capacity of a system to return to its original state following a disturbance. Tony Hodgson's typology of resilience labels these, respectively, as Type 1 and Type 2 resilience.[133] Both may be at odds with resilience theory's more dynamic conceptualisations, which emphasise change through systemic learning and/or the possibility of transformation: Types 3 and 4 in Hodgson's scheme. A system that either resists changing at all or seeks only to return to a pre-determined target state limits it own

130   Smith, A. and G. Seyfang, 2007. Grassroots innovations for sustainable development: towards a new research and policy agenda. *Environmental Politics* 16(4): 584-603. Seyfang, G., 2009. *The New Economics of Sustainable Consumption.* London: Palgrave.

131   Berkes, F. and C. Folke. 2002. Back to the future: ecosystem dynamics and local knowledge. Pp. 121-146 in Gunderson and Holling (eds.) *Panarchy.* Washington DC: Island Press.

132   Gallopín, G.C., 2002. Planning for Resilience: Scenarios, Surprises and Branch Points. Pp. 361-392 in Gunderson and Holling (eds.) *Panarchy.* Washington DC: Island Press.

133   Hodgson, A., 2010. Transformative Resilience. *http://bit.ly/2hpOQVF*

capacities for adaptation, reorganisation, learning and evolution.[134] More recent work in this area has moved beyond this to more dynamic concepts, more consistent with resilience theory, with greater emphasis on ongoing developmental process that responses to crises both reflect and support.[135]

Evidence from many different fields supports the observation that instantaneous responses to disturbance, and resilience, are intimately related to adaptability and transformability. Resilience in indigenous resource use is often the product of systemic learning from past crisis events.[136] There is evidence, particularly from research involving Arctic peoples, that existing mechanisms for coping with extreme weather events in the short-term may be the basis for long-term adaptation to climate change.[137] UK government mechanisms for disaster and crisis management show evidence of adaptive and incremental development based on learning from experience at institutional levels.[138] Many other historical examples exist where reorganisation and renewal following natural disasters has increased resilience to future disturbances, when responses have increased adaptive capacity in both infrastructure and associated management institutions.[139] This notion of resilience as an ongoing, developing condition, although perhaps most obviously manifest in crises, is consistent with the Transition idea of a property we should seek to build through community-level responses to peak oil and climate change.[140] It also fits the reconceptualisation of climate change as a wicked problem, lacking any solution as such and helping shape the context within which consideration of all other issues is framed.[141]

The inherently conservative implications of Type 1 and Type 2 resilience mean they conveniently fit discourses that seek to normalise and hence perpetuate existing imbalances of power.[142] In particularly, neoliberal rhetoric increasingly draws upon concepts of resilience both as a justifying principle, and a device for transferring responsibility for environmental and social

134   These features fit the characterisation of addictive organisations in Schaef, A. W. & D. Fassel, 1988. *The Addictive Organisation*. San Fransisco: Harper and Row.

135   Brown, K., & E. Westaway, 2011. Agency, capacity, and resilience to environmental change: lessons from human development, well-being, and disasters. *Annual Review of Environment and Resources* 36: 321-342.

136   Grove, R.H., 1997. *Ecology, Climate and Empire*. Cambridge: White Horse Press. Berkes, F., 2008. *Sacred ecology: traditional ecological knowledge and resource management*. Second edition, revised. London: Routledge.

137   Berkes, F., 2008. *Sacred ecology: traditional ecological knowledge and resource management*. Second edition, revised. London: Routledge. Pp. 161-180.

138   Rogers, P., 2011. *Comparative Approaches to Resilience*. Paper presented at the conference Resilience for Future Energy Systems, Newcastle Civic Centre, Northumbria University, 24th October 2011.

139   Gunderson, L., 2010. Ecological and human community resilience in response to natural disasters. *Ecology and Society* 15(2): 18.

140   Hopkins, R., 2010. *Localisation and resilience at the local level: the case of Transition Town Totnes (Devon, UK)*. PhD thesis, Plymouth University. Pp 72-74.

141   Hulme, M., 2009. *Why we disagree about climate change: understanding controversy, inaction and opportunity*. Cambridge University Press.

142   Neocleous, M., 2013. Resisting Resilience. *Radical Philosophy 178*. http://www.radicalphilosophy.com/commentary/resisting-resilience. Accessed November 27th 2013.

Figure 3.2.3 – *Beach art, Baltic Sea. Credit: Gesa Maschkowski.*

damage from their perpetrators to their victims.[143] In equating resilience with persistence, such discourses ignore the complexities and nuances of resilience theory in particular, most notably its attention to transformation as a vital component of resilience. Blanket assertions that all uses of the term resilience pander to these agendas are equally simplistic. There are, nonetheless, sound arguments that apolitical usage of the term resilience may tacitly and inadvertently privilege existing patterns of social relationships, which in most cases exhibit marked power imbalances, and consequently permit definitions and conceptualisations of resilience to be imposed in a top-down fashion by powerful actors.[144] Failure on the part of grassroots movements for resilience building to challenge these hegemonic notions of resilience - either explicitly or at least by making clear how, by whom, in whose interests and on what ethical premises resilience is to be defined – creates real dangers of appropriation.[145] Such appropriation would reduce personal and community resilience to, respectively,

143 Joseph, J., 2013. Resilience as embedded neoliberalism: a governmentality approach. *Resilience* 1(1): 38-52.

144 MacKinnon, D., & K. D. Derickson, 2013. From resilience to resourcefulness: A critique of resilience policy and activism. *Progress in Human Geography* 37(2): 253-270.

145 Brown, K., 2014. Global environmental change I: A social turn for resilience? *Progress in Human Geography* 38(1): 107-117.

Figure 3.2.4 – *Bike Workshop, Copenhagen. Credit: Gesa Maschkowski.*

individual and social capacities to endure the externalised costs of rampant profiteering; in its extreme form the 'Disaster Capitalism' described by Naomi Klein.[146] The next two sections consider more sophisticated approaches to both these concepts. This completes the basis for an overtly politicised notion of resilience for Transition, taken up in the final section.

3.2.4. Personal Resilience

Attention to personal resilience is prominent in Transition through its emphasis on Inner Transition (sometimes referred to as 'heart and soul'): the personal and psychological challenges that come with accepting major environmental and social change and taking active responsibility for addressing them.[147] In academic literatures, this has followed a similar trajectory to the disaster relief literature: from simplistic emphases on Type 1 and Type 2 resilience ('bouncing back') to more situated, ecological approaches which pay attention to contextual factors.[148] These more sophisticated understandings fit with practical efforts to ensure that

146 Klein, N., 2007. *The Shock Doctrine: The rise of disaster capitalism*. London: Macmillan.

147 Johnstone, C., 2006. *Find your power*. Nicholas Brealey Publishing. Macy, J., & C. Johnstone, 2012. *Active hope: How to face the mess we're in without going crazy*. New World Library.

148 Brown & Westaway *op. cit.*: 326-330.

Figure 3.2.5 – *Repair Café, Transition Bonn. Credit: Gesa Maschkowski.*

Transition creates salutogenetic environments, conducive to these personal challenges, as Maschkowski and colleagues describe in the previous chapter in this volume..[149]

It remains obscure whether any discernable patterns can be identified in the relationships between individual and contextual dimensions of personal resilience. One study (not specifically about Transition) suggests that people involved in various forms of political activism tend to have high levels of wellbeing and personal resilience.[150] However, data on links between wellbeing and active involvement in Transition Town Totnes corroborate this only partially, and equivocally.[151] A wider survey of activists in established Transition initiatives showed many were motivated by a sense of personal disconnection resulting from their

149 Also see Henfrey, T., 2014. Edge, Empowerment and Sustainability: Para-Academic Practice as Applied Permaculture Design. In *The Para-Academic Handbook: A Toolkit for making-learning-creating-acting.* London: HammerOn Press. *https://www.hammeronpress.net/shop/books/the-para-academic-handbook/*

150 Klar, M. and T. Kasser, 2009. Some Benefits of Being an Activist: Measuring Activism and Its Role in Psychological Well-Being. *Political Psychology* 30(5): 755-777.

151 Hopkins, R., 2010. *Localisation and resilience at the local level: the case of Transition Town Totnes (Devon, UK).* PhD thesis, Plymouth University. Pp 307-10.

Figure 3.2.6 – *Real World Economics Workshop on the Max-Neef Model Run by Inez Aponte and Jay Tompt, Bonn, 2014. Credit: Gesa Maschkowski.*

dissatisfaction with the current state of society, and saw Transition as a vehicle for overcoming this.[152] Alistair McIntosh considers 'cultural resilience' arising from supportive community structures essential for promoting personal resilience and allowing effective grassroots action.[153] Cultural innovations of the types documented at road protest camps,[154] and in protest cultures more generally,[155] may be examples of mechanisms for achieving this. Hodgson's typology may be relevant here: personal resilience of Type 1 (putting up) and Type 2 (recovery) intuitively seem to resonate less with Transition than Type 3 (adaptation through ongoing learning) and Type 4 (openness to, or indeed encouragement of, personal transformation); all may have contributions to make to creating salutogenetic environments.

Whatever their relationship with individual conditions, there is evidence that these health benefits can extend beyond the Transition group itself. A Health Impact Assessment of the Transition Together and Transition Streets projects in Totnes, Devon (England) identified positive effects on lifestyle, social environment, and physical environment, all of which reliably

152 Haxeltine, A., and G. Seyfang, 2009. *Transitions for the People: theory and practice of 'Transition' and 'Resilience' in the UK's Transition movement*. Tyndall Centre for Climate Change Research Working Paper 134. Pp. 19-20.

153 McIntosh, A., 2004. *Soil and Soul: people versus corporate power*. London: Aurum. McIntosh, A., 2008. *Rekindling Community*. Schumacher Briefing No. 15. Totnes: Green Books.

154 Butler, B., 1996. The tree, the tower, and the shaman: the material culture of resistance of the No M11 Link Roads Protest of Wanstead and Leytonstone. *Journal of Material Culture* 1(3): 337-363. Reprinted in Harvey, G. (ed.), 2003. *Shamanism: A Reader*. London: Routledge. Letcher, A., 2001. The scouring of the shire: fairies, trolls and pixies in eco-protest culture. *Folklore* 111(2): 147-161.

155 McKay, G., 1996. *Senseless acts of beauty: cultures of resistance since the sixties*. London: Verso.

correlate with improvements in health and wellbeing.[156] Many of Transition's key concerns – environmental degradation, climate change, ability to maintain provision of essential services in the face of declining availability and affordability of energy, and the equity implications of the uneven distribution of impacts of climatic and economic instability – are also major public health issues, and there are huge potential synergies with centralised public health initiatives that take a systemic rather than responsive approach.[157]

Human Scale Development framework developed by Manfred Max-Neef and colleagues rests on the important distinction between needs and satisfiers. The framework identifies several categories of fundamental human needs: subsistence, protection, affection, understanding, participation, recreation (in the sense of leisure, time to reflect, or idleness), creation, identity and freedom.

A possible operational link between the individual and social-ecological dimensions of personal resilience lies in the Human Scale Development framework developed by Manfred Max-Neef and colleagues.[158] This rests on the important distinction between needs – the basic requirements for human survival and flourishing, assumed to be universal – and satisfiers – the many and varied ways in which these needs can be met. The framework identifies several categories of fundamental human needs: subsistence, protection, affection, understanding, participation, recreation (in the sense of leisure, time to reflect, or idleness), creation, identity and freedom. Each of these encompasses four existential categories: being, having, doing and interacting. These two dimensions generate a 36-cell matrix in which different kinds of satisfiers can be placed:[159]

156   Richardson, J., A. Nichols, & T. Henry, 2012. Do transition towns have the potential to promote health and well-being? A health impact assessment of a transition town initiative. *Public Health* 126(11): 982-989.

157   Poland, B., M. Dooris, & R. Haluza-Delay, 2011. Securing 'supportive environments' for health in the face of ecosystem collapse: meeting the triple threat with a sociology of creative transformation. *Health Promotion International* 26(suppl. 2): ii202-ii215.

158   Max-Neef, M., A. Elizald, and M. Hopenhayn, 1989. *Human Scale Development. Conception, Application, and further reflections*. New York and London: Apex. Pp. 8.

159   *Ibid.*, pp. 32-33.

Table 3.2.1 – *The Human Scale Development Framework.*

FUNDAMENTAL HUMAN NEEDS	BEING (QUALITIES)	HAVING (THINGS)	DOING (ACTIONS)	INTERACTING (SETTINGS)
Subsistence	physical & mental health	food, shelter, work	feed, clothe, rest, work	living environment, social setting
Protection	care, adaptability, autonomy	social security, health systems, work	cooperate, plan, take care of, help	social environment, dwelling
Affection	respect, sense of humour, generosity, sensuality	friendships, family, relationships with nature	share, take care of, make love, express emotions	privacy, intimate spaces of togetherness
Understanding	Critical capacity, curiosity, intuition	literature, teachers, policies, educational	analyse, study, meditate, investigate,	schools, families, universities, communities,
Participation	receptiveness, dedication, sense of humour	responsibilities, duties, work, rights	cooperate, dissent, express opinions	associations, parties, churches, neighbourhoods
Leisure	imagination, tranquility, spontaneity	games, parties, peace of mind	day-dream, remember, relax, have fun	landscapes, intimate spaces, places to be alone
Creation	imagination, boldness, inventiveness, curiosity	abilities, skills, work, techniques	invent, build, design, work, compose, interpret	spaces for expression, workshops, audiences
Identity	sense of belonging, self-esteem, consistency	language, religions, work, customs, values, norms	get to know oneself, grow, commit oneself	places one belongs to, everyday settings
Freedom	autonomy, passion, self-esteem, open-mindedness	equal rights	dissent, choose, run risks, develop awareness	anywhere

The framework identifies five types of satisfier, which differ according to their relationship with needs:[160]

1. Destroyers – those that although directed to meeting a particular need, prejudice the long-term possibilities for its satisfaction, as well as jeopardising other needs. An obvious and topical example is continued reliance on conversion of fossil fuel and nuclear energy for the provision of energy services, violating other needs in many different ways and rein forcing lock-ins and path dependencies that undermine long-term energy security.[161]

2. Pseudo-satisfiers - that give a false impression of meeting a particular need. Much frivolous consumption can be understood as pseudosatisfaction of affective or existential needs.[162]

3. Inhibiting satisfiers – that oversatisfy a particular need and thus undermine the possibility that others are met: perhaps a defining feature of the condition of saturation with material goods that has come to be known as Affluenza.[163]

4. Singular satisfiers: that satisfy a single need and are neutral with respect to others.

5. Synergic satisfiers: that simultaneously stimulate and contribute to satisfying multiple needs. A good example might be a community food growing project, that can help satisfy needs for participation, connection, leisure, belonging, exercise and many others, at the same time as providing sustenance.[164]

This framework has some limitations: whether the needs described are universal remains to be demonstrated, and it doesn't encompass relationships between people and the natural world. Speculatively, however, it can potentially deepen our understanding of what resilience means, in the context of Transition, in various ways. Intuitively, the convergence of material and subjective factors in the relationship of needs and satisfiers feels like it fits well with Transitions attention to the balance between inner and outer work. Ability to sustain delivery of satisfiers might well provide a working definition of resilience.[165] Discrimination among different types of satisfiers can also allow evaluation of competing proposals for what

---

160 *Ibid.*, pp. 31-34.

161 Legget, J., 2013. *The Energy of Nations: Risk blindness and the road to renaissance.* London: Routledge.

162 Jackson, T., 2006. Consuming Paradise? Towards a social and cultural psychology of sustainable consumption. 367-395 in Jackson, T. (ed.) *The Earthscan Reader in Sustainable Consumption.* London: Earthscan.

163 Hamilton, C. & R. Denniss, 2005. *Affluenza: when too much is never enough.* Crows Nest: Allen & Unwin.

164 E.g. see Kneafsey, M., R. Cox, L. Holloway, E. Dowler, L. Venn & H. Tuomainen, 2008. *Reconnecting Producers, Consumers and Food.* Oxford: Berg.

165 For more informationon on possible applications of Human Scale Development to practical efforts at economic relocalisation, see Tompt, J., 2014. Relocalisation: does it meet your needs? *Stir Magazine* 5 (Spring 2014): 24-27, and *http://wellandgoodproject.wordpress.com/.*

constitutes resilience: those that emphasise destroyers, pseudo-satisfiers and inhibiting sat-isfiers immediately appear less credible that those that seek to create synergic satisfiers. By extension, a system whose putative resilience relies on undermining resilience (or assuming ever-increasing levels of Type 1 and Type 2 resilience) in interacting systems or its own constit-uent sub-systems is likely, on closer examination, to turn out not to be resilient after all.

The concept of synergic satisfiers may also help to operationalise resilience. The multifunc-tional nature of synergic satisfiers implies redundancy in capacities to meet particular needs. They also provide multiple possibilities for reorganisation and restructuring of flows of matter, energy and information, whether in response to disturbance or within ongoing processes of learning and development. More specifically, and in relation to salutogenesis, we can tentatively define desirable synergy in relation to Eric Fromm's distinction between existential dependency on having – not just ownership and control of material objects, but the instru-mentalisation of all human and other relationships – and being.[166] This is consistent with Sophy Banks' work for Transition Network on Inner Transition, which seeks to emphasise rela-tional aspects of human existence. These include relationships with ourselves as experienced through thought, feeling, embodiment and their interactions; interpersonal relationships, group dynamics, and their effects on team building and successful collaboration; and relation-ships with the natural world. If we adopt this as a basic social goal, sustaining the ability of our social-ecological system to support modes of existence based on being and the subjective, social and ecological conditions this implies becomes a key defining criterion for resilience. It seems likely that resilience, defined in this terms, will be best supported by synergic satisfiers whose functions include meeting needs related to being in whatever category. In these terms, a resilient community would be one with an existential focus on being, supported by access to multiple synergic satisfiers.

3.2.5. Community Resilience

Until recently, there was no consistent definition or characterisation of community resilience in either academic or grey literatures. While the concept of resilience was increasingly com-monly used in community development as the first decade of the 21st century unfolded, these tended to be *ad hoc* approaches that drew inconsistently upon a range of theoretical perspec-tives, and did not build on each other in any systematic, cumulative fashion. They nonetheless contributed some useful practical insights, and show a similar trajectory towards more holistic perspectives as the theoretical approaches in disaster management and psychological development on which they draw. Recent approaches, for example emphasise resilience as a state of preparedness for unexpected events that depends on ongoing activity to build and maintain capacities for community responses.[167] An important practical guide to community resilience produced by the Carnegie Foundation notes that community resilience, as a 'wicked

166 Fromm, E., 1995. *The Essential Fromm: life between having and being.* Ed. R. Funk. London: Constable.

167 e.g. Edwards, C., 2009. Resilient Nation. DEMOS. *https://www.demos.co.uk/files/Resilient_Nation_-_web-1.pdf*; UK Cabinet Office, 2016. Community Resilience Framework for Practitioners. *https://www.gov.uk/government/publications/community-resilience-framework-for-practitioners*.

issue', requires a flexible approach that allows integration of multiple perspectives. It likens resilience to a muscle, in that it is developed through ongoing community activity as a means of building the social capital that will allow the community to mobilise in response to a crisis.[168]

Recent work has attempted more systematically to integrate relevant insights from social-ecological systems on the one hand, and disaster management, community development and psychological development on the other, to move towards a consistent scientific approach to community resilience.[169] As summarised in Helen Ross's contribution to this volume, this synthesis identifies key qualities that promote community resilience: capacity for self-organisation deriving from positive outlook, suitable infrastructure, economic diversity and innovation,

Carnegie Foundation likens resilience to a muscle, in that it is developed through ongoing community activity as a means of building the social capital that will allow the community to mobilise in response to a crisis.

relationship to place; and the capacity to exercise agency through appropriate forms of leadership, suitable knowledge and skills and means to develop these further through learning, appropriate values and beliefs, engaged governance, and the ability to mobilise collectively through social networks.

While Berkes and Ross tend to treat community as a specific organisational and analytical level within the panarchy, many of the same insights apply to a broader and more flexible view. Emergent within Transition and other social movements over the years is an experience of community not necessarily as a specific geographically localised group, but as a modality of interactions, of developing and mobilising alliances and other forms of relationships around common interests and understandings. These are often manifest at local levels, within a community of place, as a Transition initiative conventionally operates. They may also arise at higher levels or across scales, linking actors with specific interests or skill sets. Examples of these include the network of national Transition Hubs,[170] ECOLISE network,[171] the Transition

168 Wilding, N., 2011. *Exploring Community Resilience in times of Rapid Change*. Dunfermline: Carnegie UK Trust. http://www.carnegieuktrust.org.uk

169 Berkes, F., & H. Ross, 2013. Community resilience: Toward an integrated approach. *Society & Natural Resources* 26(1): 5-20.

170 https://transitionnetwork.org/transition-near-me/hubs/

171 http://www.ecolise.eu

Research Network,[172] Research in Community,[173] and various individuals and groups investigating low carbon economies and livelihoods through the REconomy Project.[174] These all intersect and cross cut each other, with local Transition initiatives and other projects and national and regional networks, and with other communities of practice with common interests. How they might operate across scales is the topic of the next, concluding section.

3.2.6. Action Across Scales: Framing and Diversity

The *pathways* approach to sustainability highlights the importance of framing.[175] A 'system' is an analytical construct – implicit in any usage of resilience - that identifies a certain set of relationships (exchanges of energy, matter and/or information) as the most important features of a complex, messy reality. Different models can describe and analyse the same reality in different ways. These differences in large part depend on framing, which in relation to system definition means decisions about which features either to emphasise or to exclude.

Framing is crucial to the way this chapter defines and treats resilience: it influences which inputs, outputs and feedbacks are treated as important as well as what potential and other contextual factors are taken into consideration. In

> *Answers to the questions, 'Resilience of what, to what, and in whose interests?', depend on framing.*

other words, answers to the questions, 'Resilience of what, to what, and in whose interests?', depend on framing. As allusions to resilience in neoliberal discourse show, not all framings of resilience are as inclusive, equitable and/or sustainable as those in the Transition movement would wish. The Common Cause report highlights the importance of making framing explicit, both in order to express openly the values that support one's own position, and to reveal implicit values and hidden agendas that may lie behind those of others, particularly those in power.[176] This has two main implications for Transition. First, a need to identify where its own framings and associated actions create barriers to inclusion and so undermine efforts to build resilience. Second, to provide a means to reveal and engage with power without compromising what is necessarily a politically radical agenda.

Diversity in perspective, and hence framing, is an important feature of adaptability and resilience. The community of people co-dependent on a particular local resource base and/or

---

172 *http://www.transitionresearchnetwork.org*

173 *http://www.researchincommunity.net*

174 *http://www.reconomy.org*

175 Leach, M., I. Scoones and A. Stirling, 2010. *Dynamic Sustainabilities*. London: Earthscan. Pp. 43-52.

176 Crompton, T., 2010. *Common Cause: the case for working with our cultural values*. Godalming: World Wide Fund for Nature. *http://www.wwf.org.uk/change.*

Figure 3.2.7 – *Beds at Neuland Community-Garden, Cologne. Credit: Gesa Maschkowski.*

infrastructure may have different outlooks and priorities, and both perceive and use it in different ways. The UK's Strategic National Framework on Community Resilience, for example, stresses a need to maintain global food supply chains, a marked contrast with Transition's emphasis on reducing dependence on these.[177] Studies of traditional resource users have shown that this diversity can contribute to evolutionary resilience by keeping different parts of the system in different ecological states and maintaining a broad base of human and natural capacities through which to respond to change.[178] Recognising this, Transition's emphasis on including as wide a range of voices as possible is practical as well as ethical. It is therefore a key concern– pointed out by many commentators and widely recognised within the movement – that in practice it falls short of its aspirations to be genuinely inclusive.

177   Cabinet Office, 2011. Strategic National Framework on Community Resilience. Pp. 7-8.

178   Berkes, F. and C. Folke. 2002. Back to the future: ecosystem dynamics and local knowledge. Pp. 121-146 in Gunderson, L. and Crawford S. Holling (eds.) Panarchy. Washington DC: Island Press. Crane, T. A. 2010. Of models and meanings: cultural resilience in social–ecological systems. Ecology and Society 15(4): 19. [online] URL: http://www.ecologyandsociety.org/vol15/iss4/art19/.

Figure 3.2.8 – Beds at Allmende-Kontor Community Garden, Berlin.

Barriers to inclusion in Transition can take many forms. For many, the term 'community' can connote a sense of affluent rural England, reinforcing caricatures of Transition as appealing only to a largely white, university educated, middle class demographic.[179] A commitment to methodologies intended to be inclusive and empowering can, ironically, favour those with the confidence to express themselves in public or semi-public settings, often reflecting relative privilege of background, status, or education.[180] Transition's flexible, non-prescriptive DIY methodology offers a ready framework for addressing this.[181] Attention to concrete local issues, such as food, engages a wide range of actors and perspectives without overt or direct interest in peak oil or climate change, the original high level drivers of Transition.[182] Nonetheless, in practice becoming a truly inclusive movement remains a work in progress at all levels.

179 Aiken, G., 2012. Community Transitions to Low Carbon Futures in the Transition Towns Network (TTN). *Geography Compass* 6(2): 89-99.

180 Cohen, D., 2010. *Reaching out for resilience: exploring approaches to inclusion and diversity in the Transition movement*. M.Sc. thesis, Centre for Human Ecology.

181 Senior, L., 2011. *The Links Between Resilience, Diversity and Inequality: The View from Transition Durham*. Masters dissertation, Durham University.

182 Mycock, A., 2011. *'Local Food' Systems in County Durham: The capacities of community initiatives and local food businesses to build a more resilient local food system*. Masters dissertation, Durham University.

Differences of perspective between grassroots and top-down framings provide opportunities for synergy as well as conflict in cross-scale working. In the later stages of research towards the Totnes EDAP it became apparent that the grassroots approach to information gathering had omitted vital perspectives from business and local authority.[183] The indicators it derived differ markedly from those in a more top-down study by the New Economics Foundation,[184] indicating different - and complementary – perspectives and priorities at community level compared with those of local authorities. The Carnegie Foundation report notes that activists, professionals and policy-makers offer different and possibly complementary contributions, and sees this as a basis for collaboration.[185]

Examination of urban Transition initiatives in the UK, which have tended to converge on a tactic of connecting separate local neighbourhood initiatives across larger cities, suggests that partnership with local authorities operating citywide is a key prerequisite for progress.[186] The case studies from Peterborough, Bristol and Spain in Section Two of this volume support this finding. They also highlight the risk inherent in any form of engagement with incumbent regimes: that it will entail compromise of a type that conflicts with the ultimately radical goal of transformation to a fair, sustainable and prosperous society.

There is also evidence that such collaborations can be subversive of the status quo and the agendas of those who would seek to maintain it. Resilience can be a powerful conceptual tool for achieving such outcomes. Post-earthquake reconstruction in Canterbury, New Zealand, involved extensive collaboration between community groups, and municipal authorities and emergency services. Collaborating around the conceptual theme of resilience enabled the community groups involved to assert their own understandings of what, in practice, resilience implies, and to renegotiate the premises of local democracy through the fact and nature of their participation – or not – in key decision-making processes.[187] In the UK,

> *Examination of urban Transition initiatives in the UK suggests that partnership with local authorities operating citywide is a key prerequisite for progress.*

---

183 Hopkins, R., 2010. *Localisation and resilience at the local level: the case of Transition Town Totnes (Devon, UK).* PhD thesis, Plymouth University. P. 337.

184 Hopkins, R., 2010. *Localisation and resilience at the local level: the case of Transition Town Totnes (Devon, UK).* PhD thesis, Plymouth University. Pp 338-340

185 Wilding, N., 2011. *Exploring Community Resilience in times of Rapid Change.* Dunfermline: Carnegie UK Trust. P. 2. *http://www.carnegieuktrust.org.uk*

186 North, P., & N. Longhurst, 2013. Grassroots localisation? The scalar potential of and limits of the 'transition'approach to climate change and resource constraint. *Urban Studies* 50(7): 1423-1438.

187 Cretney, R., & S. Bond, 2014. 'Bouncing back'to capitalism? Grass-roots autonomous activism in shaping discourses of resilience and transformation following disaster. *Resilience* 2(1): 18-31.

the trend towards community ownership and governance of energy generation infrastructure, when locally articulated as part of a strategy for building resilience, directly challenges national policy measures based on elite framings of energy security.[188]

Engaging power, particularly on its own terms, is always an uncomfortable experience for those who seek to subvert it, one which both Transition and Resilience Theory have been accused of evading. A recurrent criticism of Transition is that the absence of any explicit account of power makes it politically naïve, and hence capable neither of achieving its goals in practice nor avoiding co-option by existing regimes.[189] Resilience theory, too, has been criticised for being apolitical.[190] However, a scientifically grounded resilience, that makes explicit its ethical conviction to social justice, is deeply subversive of neoliberal values.[191] The

> *Resilience, according to the definitions used here, in practice implies the replacement of centralised power structures with open and inclusive governance mechanisms.*

Open Space session on Resilience and Community Action at the Resilience 2014 conference, reported in Chapter 4.1 of the present volume, held multiple accounts from direct experience of the personal challenges involved in pushing against entrenched institutional barriers to positive transformation. This intransigence, and the discomfort involved in confronting it, reflects the power of a felt and enacted politics of resilience. When this is the basis of direct engagement with power, it may challenge it more effectively than an overtly politicised position that undermines possibilities for constructive dialogue.

Confusion over the definition, meaning and implications of resilience in many of the discourses in which it has become prominent opens up a discursive space with powerful critical possibilities.[192] Resilience, according to the definitions used here, in practice implies the replacement of

188 Butler, C., S. Darby, T. Henfrey, R. Hoggett, & N. Hole, 2013. People and Communities in Energy Security. In Mitchell, C., J. Watson & J. Whiting (eds.) *New Challenges in Energy Security: The UK in a Multipolar World.* London: Palgrave MacMillan.

189 Alloun, E. & S. Alexander, 2014. *The Transition Movement.* Simplicity Institute Report 14g. *http://www. vikalpsangam.org/static/media/uploads/Resources/transitionmovement_simplicity_inst.pdf*

190 Cretney, R., 2014. Resilience for Whom? Emerging Critical Geographies of Socioecological Resilience. *Geography Compass* 8(9): 627-640.

191 Nelson, S. H., 2014. Resilience and the neoliberal counter-revolution: from ecologies of control to production of the common. *Resilience* 2(1): 1-17.

192 Blewitt, J., & D. Tilbury, 2013. *Searching for Resilience in Sustainable Develoment: Learning Journeys in Conservation.* London: Routledge.

centralised power structures with open and inclusive governance mechanisms. It also implies transformation in systems that can persist in their current form only by continually undermining resilience elsewhere. Accordingly, a combination of robust science and explicit, unwavering ethical commitment empowers us to take control of this space and within it reveal the inherent contradictions in neoliberal discourse on resilience. This may in turn contribute to the transformation in thinking called for in Maja Göpel's chapter in the present volume.

This radical conceptual deployment of resilience also compels Transition to examine its own agenda, and to consider its relationship to political ecology in a global context. As Henfrey and Kenrick argue in Chapter 4.2, alongside other efforts to oppose the ongoing enclosure of material and cultural resources within the market realm, Transition can be seen as part of a wider global movement to defend and extend the commons.[193] In Chapter 4.3, Kuecker argues that moving from a rhetorical alignment with majority world struggles to genuine and meaningful solidarity involves ongoing critical examination of the ways in which the practice of Transition reflects its origins among relatively privileged sections of the global population. Challenging power means revealing and confronting our own power and its legacy, a necessary and desirable implication of taking action for resilience.

Acknowledgements

Initial work on this paper took place within the research project 'Connection, participation and empowerment in community-based research: the case of the Transition movement', funded by the Arts of Humanities Research Council (UK) within the Connected Communities Research Programme. We would like to express our gratitude for the financial support, our appreciation of the programme's efforts to open new frontiers in government-funded academic research, and our regret that this failed to attain its full transformative potential as a result of intransigence in established academic regimes. The content benefited greatly from conversations at workshops on Transition and Resilience held as part of this project in Bristol on May 12th 2012 and July 3rd 2013: our thanks goes to all who took part, and in particular to Sophy Banks, John Fellowes, and Gesa Maschkowski for their contributions to subsequent email dialogue.

193   Also see Kenrick, J., 2012. The Climate and the Commons. In Davey, B. (ed.), *Sharing for Survival*. Feasta. *http://www.sharingforsurvival.org/index.php/chapter-2-the-climate-and-the-commons/*

3.3. Shedding some Light on the Invisible:
the Transformative Power of Paradigm Shifts
MAJA GÖPEL

"A transition to sustainability demands profound changes in understanding, interpretative frameworks and broader cultural values, just as it requires transformations in the practices, institutions and social structures that regulate and coordinate individual behaviour. In this context, it is essential to get to the position where people, industry and governments can easily distinguish between objective facts and opinions that are presented as facts to advance particular interests, and rely on the former to make informed decisions."[194]

In order to make sense of the world humans create ideas and stories about why they are here, what the purpose of their life-journey is, and how to relate to their human and natural environment. The results are individual mindsets that lie at the heart of identities, and social paradigms that structure socio-political development processes. The latter include widely accepted common sense, canonized knowledge, and cultural narratives enveloping human role definitions and cooperation agreements. They in turn are reified: concretizations of ideas and stories that become structural or even material features of the context in which future thinking, observation and being take place. Thus, subjective ideas and intersubjective stories or narratives are intricately linked with the 'objective' world. They can be a source of vision, innovation, creativity and flourishing progress — and a source of mental barriers, strategic power or even forceful domination.

Understanding this structural-material impact of ideas and how it shapes what could be called the 'patterned freedom' of human development lies at the core of the 'reflexive' social sciences. Reflexivity is a uniquely human capacity that enables people to become aware of the biasing forces and effects of socialization and to identify where institutional path dependencies and guiding stories drive societies along development routes that are not

194 UNEP Global Environmental Outlook 5 Report, 2012, p. 447.

Figure 3.3.1 – Signpost in Buthan. Credit: Maja Göpel.

(any longer) in line with overarching goals and aspirations. Assessing the underlying assumptions and unstated ideas upon which social processes and institutions have been built, justified, maintained and adapted empowers us to break free from them if necessary. Reflexivity as an empowering and emancipating activity therefore forms a core of strategic engagement for changes to societal structures and institutions that have been set up and form 'reality' today. The World Social Sciences report 2013 coined the term 'futures literacy' when discussing leverage points for deeper and wider system changes for sustainable futures:

> *"The complexity of these processes of transformation raises a number of questions, most notably about people's capacity to imagine futures that are not based on hidden, unexamined and sometimes flawed assumptions about present and past systems. 'Futures literacy' offers an approach that systematically exposes such blind spots, allowing us to experiment with novel frames for imagining the unknowable future, and on that basis, enabling us to critically reassess actions designed in the present."*
> (World Social Sciences Report 2013, p.8)

This paper argues that uprooting some of the hidden, unexamined and quite flawed assumptions created by neoclassical economics and its embedding in culture and social institutions is a transformative leverage point for translating sustainability visions and commitments into reality. Meanwhile, futures literacy also means exposing where potentially radical visions and ideas for future developments are successfully re-framed and co-opted into 'old' neoclassical paradigm and development patterns, dissolving their potentially transformative edge. Sustainable development is the main example discussed here; as discussed elsewhere in this collection, the same is increasingly true for resilience.

The structure of the paper is as follows: a brief introduction to the original definition of sustainable development in 1992 precedes a rough outline of why the neoclassical economic paradigm cannot provide any meaningful insight into how this agreed goal can be achieved. Tying this analysis back with research on transitions in complex systems it then combines the multi-level-perspective (MLP) on societal change with Gramscian hegemony theory on leading with least resistance to argue that replacing the neoclassical mindset or paradigm is a high leverage point for system transformations towards sustainability. The outlook briefly summarizes some approaches from alternative sustainability economy movements that tackle the identified blind spots in neoclassical economics head-on. In a first assessment they show a surprisingly high degree of commonality in their ideas, stories and governance solutions that could shape into a new economic paradigm and mindset suitable to coordinate diversified initiatives into a political movement or Gramscian 'common will' for structural change.

3.3.1. Sustainable Development: Which Vision and Goals?

In 1987 the United Nations appointed the World Commission on Environment and Development (WCED) and published its path breaking report on sustainable development. It highlighted how the 20th Century's path of economic development had heavily damaged nature while still keeping a majority of people in poverty. The call was to replace this path with 'Sustainable Development' development which, "[M]eets the needs of the present without compromising the ability of future generations to meet their own needs."[195] To specify this new vision, the report highlighted two key points for attention and intervention. These

> *Sustainable development has not been achieved due to the fact that the overarching development narrative and its underlying neo-classical economic assumptions were not seriously challenged.*

were "[T]he concept of 'needs' in particular the essential needs of the world's poor," to whom, it argued, "overriding priority should be given," and, "the idea of limitations imposed by the state of technology and social organization on the environment's ability to meet present and future needs."

In 1992 the Rio Declaration of the United Nations made sustainable development the overarching policy principle of international cooperation. Today's statistics provide ample evidence that sustainable development has not been achieved. I will argue here that this is at least partly due to the fact that the overarching development narrative and its underlying neoclassical economic assumptions were not seriously challenged. Rather than being weakened

---

195   WCED, 1987, Our Common Future, Chapter 2, Paragraph 1, online at: *http://www.un-documents.net/our-common-future.pdf*

by challenges like natural limits to growth or unfair distribution of its impacts and the links between market liberalization and increased inequality, this paradigm has shown a remarkable ability to reinforce itself by incorporating them. Historically, such co-optation was helped by the collapse of the Soviet Union around the same time, leading to declarations of "The End of History" (Francis Fukuyama). Also, powerful promoters of this storyline and paradigm are usually spared from suffering the consequences of its implementation: it provides a terrific rationale for feeling comfortable with being much better off than others.

3.3.2. Neoclassical Economic Paradigm: Which Insights for Sustainable Development?

A paradigm and its associated societal narratives rest on some core ideas or concepts. These are more than simple flashes of thought, a mere slogan or a buzzword. According to institutional political economists Morten Boas and Desmond McNeill, such an idea:

> "[H]as some reputable intellectual basis, but... may nevertheless be found vulnerable on analytical and empirical grounds. What is special about such an idea is that it is able to operate in both academia and policy domains."[196]

Providing the language and sense-making that people apply in order to govern their own existence in the world, ideas also become part of the common sense and narratives according to which collaboration is set out and institutions are designed.

Looking at what the narrative of neoclassical economics has offered to make the sustainable development vision become reality one should not be too surprised that it has not come round: in its analytical concepts or core ideas this paradigm has lumped all human needs – the one key point in the sustainable development definition - into one undifferentiated concept of 'utility maximization'. It is regarded as a fundamental universal law or human condition that actors selfishly, insatiably and rationally pursue the never-ending maximization of pleasure. All other concepts and explanations stem from this core idea, and the prime source of pleasure is considered to be consumption.

'Utility maximization' is regarded as a fundamental universal law or human condition that actors selfishly, insatiably and rationally pursue the never-ending maximization of pleasure, and the prime source of pleasure is considered to be consumption.

---

196   Desmond McNeill and Morten Boas, 2004, Global Institutions and Development. Framing the World? London: Routledge, p. 1.

Unsurprisingly, according to neoclassical economics, the natural best development path for these 'representative actors' is one of eternal growth in consumption and therefore production: utility and needs satisfaction, or happiness, will then keep on rising forever as well. The most efficient and just institution to bring selfish actors into the cooperation necessary for production and exchange processes are markets in which supply and demand are matched through prices: 'freely' negotiating actors end up signing 'voluntary' contracts in which everyone attempts to get the best cost-benefit deal for him- or herself.

Each price therefore indicates 'willingness to pay' and provides a handy indicator of the utility created by a particular good or service: selfish people only pay as much as they really value something. Taken to the level of explaining entire economies and their development we find the origin of the third universal law, that of equilibrating competitive markets: every product finds a buyer once the price is right and human needs will therefore steer production in the right direction.

Figure 3.3.2 – *Traffic Jam. Credit: Gesa Maschkowski.*

Since this paradigm looks primarily at the point of decision-making but excludes the context in which it takes place (e.g. the distribution of what people can offer in the 'free agreements', what the content of these capacities are, or which ends they serve) it includes no concept of power. In the socio-cultural domain of sense-making it easily translates into a discourse that particularly benefits comparatively wealthy groups or individuals: market competition drives people to highest performance and thus the revenue a particular skill generates in the market is a just expression of the social value of this person's contribution. If people do not manage to get a return for their offerings it is their own deficiency rather than anything else: they did not try hard enough to offer something desirable, did not provide solutions for the needs of people.

In these equations it is possible to completely eradicate other forms of life as long as overall economic output keeps on rising: environmental and social damage remain invisible as long as they do not register in market prices and what is actually traded remains invisible in this price-based utility measurement.

With reference to respecting the laws of nature and their ability to sustain satisfaction of needs – the second key point of the sustainable development definition – the equilibrating market will also do the trick. Once increasing scarcity in natural resources drives up their prices, smart creative entrepreneurs will come up with alternatives to generate the same consumption options from cheaper sources or with different technologies. The way the environment was 'integrated' with economics (a central demand from the Brundtland report) was to internalize nature into abstracted cost-benefit equations. The solution was found by including prices of resource units in calculations of production costs and therefore steer usage in a sustainable direction – or rather, in a direction in which resource destruction does not compromise human consumption.

In these equations it is possible to completely eradicate other forms of life as long as overall economic output keeps on rising: environmental and social damage remain invisible as long as they do not register in market prices and what is actually traded remains invisible in this price-based utility measurement. Only money flows or 'exchange values' register in the calculations and the concept of 'capital substitutability' explains that this monetary wealth can be transformed into anything else – including somehow uncomplicated environmental restoration should humans really not succeed in finding useful substitutes for certain resources. Since therewith no limits to economic growth exist, socially everything is fine as long as some money is invested in education so that people become more successful in selling their value in the markets.

Figure 3.3.3 – Trash at the Beach. Credit: Gesa Maschkowski.

What this paradigm therefore leaves unanswered is the core of the sustainable development vision:

› How can we prioritize the needs of the poorest in a meaningful way if we do not differentiate use value of goods and services for healthy existence for all from the expression of exchange value in pricing mechanisms that are silent about what they stand for? Expressions of increased wealth like housing bubbles, astronomical expenditure on abstract pieces of art and fun rides to Mars have the same apparent 'utility gain' as providing food, shelter and healthcare to the poorest.
› How do we know if we risk overexploiting even renewable resources if we only look at the flows that register in markets and not at remaining stocks and the complex dynamics of their reproduction? Every alternative solution needs resources and transmission structures as long as humans are not directly converting solar energy into all they need for survival.

The bottom line then is the question of how we understand and meet human needs or devise strategies for aligning satisfaction strategies with the limits of a finite planet on a long-term basis within such a theoretical framework. One in which individuals cannot stop wanting ever more even if they are plump, filthy rich and burnt out and where natural life cycles outside of production functions do not exist.

3.3.3. The Role of Mindsets and Paradigm Shifts in Social Transformations

According to theories based on reflexivity, human decision-making processes are guided by an individual's worldview, mindset or consciousness. In making choices, he or she in turn influences the sense-making of counterparts and observers. Thereby, social groups continually co-create their living conditions. Humans are both the subjects and the objects of history, as political economist Robin Hahnel points out:

> "[E]very person has natural attributes similar to those of other animals, and species char-
> acteristics shared only with other humans - both of which can be thought of as genetically
> 'wired-in.' Based on these genetic potentials people develop more specific derived needs
> and capacities as a result of their particular life experiences. While our natural and species
> needs and power are the results of past human evolution and are not subject to modifica-
> tion by individual or social activity, our derived needs and powers are subject to modifica-
> tion by individual activity and are very dependent on our social environment."[197]

This dependence involves significant limitations on the ability of single people to change social roles defined by society's major institutions within which most of our activity takes place. This is one of the main causes of inertia in bigger organizations and societies. Social scientists, transition researchers and political economists use the concepts 'paradigm shifts', 'path dependencies' and 'hegemony' to assess these processes in more detail.

The term 'paradigm shift' originates from the philosophy of science and usually references Thomas Kuhn as the original thinker in this context. In his 1962 book *The Structure of Scientific Revolutions* he wanted to describe a change in the thought patterns and basic assumptions with which scientific analyses are addressed. In scientific terms, paradigms comprise assumptions that are epistemological (what can we know), ontological (what can be said to exist and how do we group it), and methodological (which guiding framework for solving a problem is suitable). In the context of worldviews many add axiological aspects (which values are adopted). Depending on how these are defined, one and the same event will be interpreted very differently. Kuhn examined how the standard definitions of these assumptions determine *which* questions will be raised when assessing a certain issue, *how* they will be raised, *what* will be observed and *how* these results will be interpreted. Usually, competing paradigms hold different assumptions and therefore one and the same event will be analyzed differently and proposed solutions to the same problem will vary significantly, depending on assumptions about actor behavior, processes of development and system characteristics.[198]

Generally, the existence of competing paradigms already prohibits the declaration of full objectivity or the existence of unshakable truth. This is particularly true for social sciences like economics where the ideas about the world inform the institutions we build to govern the world and therefore how the future world or 'reality' looks like. Thus, as Kuhn claimed,

197　Robin Hahnel, 2002, The ABCs of Political Economy. A Modern Approach, London: Pluto Press, pp. 4-5.

198　Thomas Kuhn, 1962, The Structure of Scientific Revolutions. University of Chicago Press.

what is considered to be 'true' in science has the quality of a consensus within the scientific community. Since the people forming this consensus have undergone processes of socialization themselves, science is never completely free of the mindsets that those involved bring to the table or laboratory. During strong dominance of one particular paradigm like that of neoclassical economics, however, research results not conforming to the paradigm's prediction are usually interpreted as a mistake by the researcher or dismissed as not worthy of further inquiry instead of a falsification of the paradigm's assumptions. When

> *Generally, the existence of competing paradigms already prohibits the declaration of full objectivity or the existence of unshakable truth.*

paradigms shift, however, new ways of interpretation and understanding that formerly would not have been considered valid are opened up and new truth claims can emerge.

Neoclassical economics has a long tradition of defending its foundational 'natural laws' of human behavior with vague amendments like 'less-than-perfect' information or 'bounded rationality' in decision-making, but has never gone through a real ontological shift. The socio-economic concept of path dependency sheds some light on why this is understandable. It explains why social institutions carry a self-stabilizing momentum fostering the continuation of the status quo. If the status quo is challenged, it translates into a deviation from the 'normal' way of doing things. Informal rules and routines in organizations tend to render such deviations as less easily acceptable or adaptable. They challenge beliefs, create fear of loss through role changes and include higher transaction costs since established processes are changed. In addition, institutionalization and the creation of manufactured infrastructures lead to material-technological lock-ins that are truly difficult to change even if people decide that an alternative way of providing, for example, public transport or energy would now be better.

Meanwhile, being socio-political actors, individuals or groups who are struggling for change will defend proposed alternatives rationally, seeking to justify them not only to themselves but also to others whose support they wish to gain. The more they manage to appeal to widely established convictions and canonized knowledge, cultural narratives, belief systems and the 'derived needs' in a particular group, the more likely their particular solution is to find supporters. Thus, proponents of status quo solutions and their path dependencies and the social roles, vested interests and structural procedures embedded therein, always have an advantage over those with new proposals. The prevailing ethics, norms, rules and laws in the given context and the distribution of skills and power to navigate these, effectively provide a framework for action that is a biasing yet rather 'invisible' source of justification and legitimization in political struggles.

Institutionalized ideas function as much as path dependencies as do technological and material infrastructures or economic cost-benefit calculations. They are part of the structural power of the status quo against alternative ideas and proposals.

In order to capture the effects of this framework for action and the role that the 'mental glue' of paradigms and mindsets play in defense of the status quo, political economist Antonio Gramsci developed the concept of hegemony in the 1930s. He coined this term because he wanted to find explanations for situations in which we observe a few enjoying far more wealth and freedom than the majority despite living in democracies with presumably similar rights of citizenship. The concept draws attention to the role of culture and social norms in securing leadership and the resilience of particular governance solutions. It also highlights how strategic use of science or cultural framing can dress up particular political positions. Gramsci's work draws heavily on that of Machiavelli, namely *The Prince*. For successful leadership towards the founding of a

Figure 3.3.4 – Bicycle Stands in the Tram, Copenhagen. Credit: Gesa Maschkowski.

new state, this work proposes the use of a 'dual perspective' of consensus and coercion. A central role in it is to offer a narrative on what this society and living in it are about, and which policies and programs are therefore in the common interest.[199]

This narrative has the quality of a 'social myth', "[A] political ideology expressed neither in the form of a cold utopia nor as learned theorizing, but rather by a creation of concrete fantasy which acts on a dispersed and shattered people to arouse and organize its collective will."[200] The 'collective will' for Gramsci is a group of people strategically promoting the ideas and stories supporting the social myth so that over time heterogeneous interests are welded together under a single aim, on the basis of an equal and common conception of the world.

The social myth at the center of this common conception therefore plays a very important role in legitimizing or justifying the adequacy of the specific norms, practices, institutions and regulations put forward or in place. Having become the dominant common sense in this society, it overlies individual sense-making and influences the development of attitudes as to why we should behave in a certain way or expect others to do so.

Such institutionalized ideas function as much as path dependencies as do technological and material infrastructures or economic cost-benefit calculations. They are part of the structural power of the status quo against alternative ideas and proposals. In reflexive science this is widely acknowledged even though the degree to which scholars and practitioners under-stand narratives or ideologies as a strategic mechanism of the elite to lead with least resist-ance will already depend on the paradigm. Some are closer to calling changes in perception of the world 'learning' and 'evolution' whilst others will enunciate the power aspects and expose elements of 'domination' in standardization and setting collective rules.

Regardless of an individual's position on this spectrum, most will agree that without a good narrative and some empirical examples of why changing the status quo is actually more in the interest of powerful players or the general good, it is very difficult meaningfully to change existing institutions and development pathways without full-blown crises that threaten their perpetuation. Preemptive adaptation or transformation strategies therefore rest on ideas and visions, or mind and paradigm shifts that redefine the understanding of what are possible solutions in a given situation, or even the imaginary of potential future lives, socio-economic set-ups and human-nature relations. After all, it is human sense-making and engagement that drive socio-economic and political developments and find materialization in the institutions that constitute the 'reality' of today.

Social system scholars like Donella Meadows therefore analyze paradigms as the 'source of systems', informing the purpose on which these are set out to deliver. In system transforma-tion strategies, paradigms therefore rank as the second highest leverage point, above rule

199   Antonio Gramsci, 1971, Selections from the Prison Notebooks, International Publishers, p.126
200   Ibid.

changes and any other standards or metrics: "The shared idea in the minds of society, the great big unstated assumptions - unstated because unnecessary to state; everyone already knows them - constitute that society's paradigm, or deepest set of ideas about how the world works."[201] Once these reference frameworks start changing we observe a widespread questioning of the institutions in place, the goals they serve and the processes on which they rest. People begin to ask, "What is the purpose here?" From a Gramscian point of view one would say that the hegemony of particular ideas or narratives and therefore their legitimizing power are challenged. Coupled with frictions in the economic-technological reproduction circuits, conditions emerge for a 'structural crisis' that holds the potential for more radical system change.

3.3.4. Embedding Mindset and Paradigm Shifts in Transition Theory

One rather recent research discipline seeking to understand and conceptualize wider and deeper system change is transition theory.[202] One of the central concepts in this research community was developed by Frank Geels and is referred to as the Multi-Level Perspective (MLP).[203] It draws on structuration theory in sociology and distinguishes qualitatively different organisational levels in societies according to their degrees of changeability and resistance to change. This does not imply a hierarchical structure (change can arise at any level) but does express how changes at overarching levels typically impact path dependencies that structure the embedded ones. It distinguishes three such levels:

› a *niche level* where experiments or pioneering innovations are undertaken by small units or 'situated groups' that can change fastest and deviate most from the established framework for action because they show few interdependencies with overarching or neighboring systems,

› a *regime level* whose structures include well-established practices, rules, science and technologies that govern social interaction on the societal level and, through institutional settings and feedback loops, tend to stabilize the status quo,

› an overarching *landscape level* of slowly changing, rather exogenous development trajectories like environmental conditions, major infrastructure, deeply anchored economic institutions like the market system, and worldviews or social values. These form the backdrop for lower level developments.[204]

201 Donella Meadows, 2009, Thinking in Systems. A Primer, Earthscan, p. 162

202 For a website with a manifesto on this research approach, links to articles and the annual conference see *http://www.transitionsnetwork.org*

203 For an overview of joint concepts and differences between sub-schools see the book by Jan Shot et. al., 2010, Transitions to Sustainable Development. New Directions in the Study of Long Term Transformative Change, London: Routledge

204 Depending on the author you find slightly diverging descriptions on where structurations like market patterns or policy orientation rest, whether at regime or landscape level. Each case may allocate these slightly differently, depending on the actual system under consideration.

The figure below depicts the development of structurations at all these levels as the result of parallel processes in diverse subsystems influencing each other and reacting to changes or shocks in connected or overarching structurations. The landscape level is impossible to change purposefully in the short term, but it can bring about shocks that lead to rapid change at regime or niche levels, like natural disasters. The changing configurations create different impulses and spaces for transformations, many changes at embedded levels also triggering reactions at overarching ones.

From this point of view, transitions to sustainable development are conceptualized as long-term multi-actor processes involving interactions among citizens and consumers, businesses and markets, policy and infrastructures, technology and cultural meaning. Resistance results from direct intervention on the part of other actors or groups with different interests or views but also from the various types of path-dependencies outlined above.

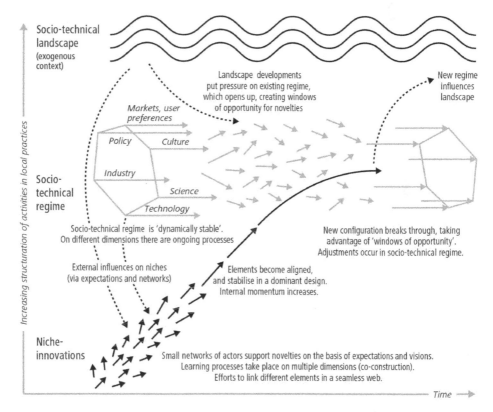

Figure 3.3.5 – *The Multi Level Perspective on System Transformation. From Göpel, M., 2016. The Great Mindshift. p. 21 (adapted from Geels, F., 2002. Technological transitions as evolutionary reconfiguration processes: a multi-level perspective and a case-study. Research Policy 31(8-9): 1263), Reproduced here with permission from M. Göpel and F. Geels.*

In their analyses, however, most transition studies on changes in socio-technical systems focus on the more visible and tangible types of path dependencies and the relationships between new technologies and social practices. The role of mindsets, narratives or cultural aspects and their potential structural power in pushing or blocking transformative change is usually not assessed. Some reference is made in a list of five 'main ingredients' for successful transformation that system innovation consultant Charles Leadbeater has put together. Taken together, these ingredients amount to what Gramsci would have called a structural crisis in a framework for action:

> *The new principles mentioned are the result of the conviction that the system could be organised differently, and ideas as to how this can happen and what purpose it could then fulfill.*

1. Failures and frustrations with the current system multiply as negative consequences become increasingly visible.

2. The landscape in which the regime operates shifts as new long-term trends emerge or sudden events drastically impact general availability or persuasiveness of particular solutions.

3. Niche alternatives start to develop and gain momentum; coalitions start forming that coalesce around the principles of a new approach.

4. New technologies energize alternative solutions, either in the form of alternative products or as new possibilities for communication and connection.

5. For far-reaching regime change rather than small adaptations and co-optation by the old regime, dissents and therefore fissures within the regime itself are key. Possibly called niches within the regime, in joining coalitions for change they will help bring the system down or at least significantly change its current set-up and development dynamic.[205]

A core functional ingredient in this sequence is of course the 'new approach' mentioned in point 3, around which multiple actors and groups coalesce. The new principles mentioned are the result of the conviction that the system could be organized differently, and ideas as to how this can happen and what purpose it could then fulfill. In the following paragraphs I

205 Charlie Leadbeater, 2013, The Need For Regime Change, in: Systemic Innovation: A Discussion Series, Nesta Foundation, pp.31-32, for download at *http://www.nesta.org.uk/publications/systemic-innovation-discussion-series*

illustrate that hegemony theory can illuminate this point and provide an informative perspective on system transformation.

Contemporary political economists like Stephan Gill have developed analytical concepts of Gramscian hegemony that fit nicely with the MLP. They highlight how the neoclassical development story finds different forms of expression at each level and help sharpen understanding of the degree to which this paradigm has been encoded into societies and culture.

At the landscape level, *market civilization* describes the overarching structure of the market system and the hegemonic narrative of competitive growth according to which all relationships should be shaped by commodification and organized according to price signals. According to Gill, this deeply anchored grammar nurtures an ahistorical, economistic and materialistic, self-oriented, short-term and ecologically myopic perspective on how the world works.[206]

The regime level is marked by what Gill calls *new constitutionalism*, describing how laws, regulations, social practices and artifacts are necessary to create commodity forms and market patterns from human skills, ecosystem services or credit relations. Their amendment and expansion transform the organizational logic of formerly non-marketized areas of life, like governance of nature in the form of Emissions Trading Systems. The most impressive of those examples may be the addition to the finance system of 'third markets' of derivatives that have no existence beyond digital numbers on a screen and legal frameworks promising their owners' claims to real resources.

Thus, from a hegemony point of view, such regime structures armor the market civilization perspective on human development via tangible manifestations in norms, rules, role definitions and infrastructures that in turn become people's experienced reality. It is of this that the edited volume by Boas/McNeill on 'framing the world', cited above, is providing research examples: how the creation of international institutions like the International Monetary Fund or World Bank is driven by players convinced of the neoclassical paradigm for globalization strategies (in its policy implications often referred to as the 'Washington Consensus') and how these in turn lead to a restructuring of the living conditions of people on the ground. In effect, living embedded in such social and institutional systems and path dependencies pushes individuals closer to behaving and organizing their own lives in accordance with the predictions and demands of the market civilization. The term 'armoring' also indicates that those interests and groups benefiting from these particular regime solutions can count on being defended with the force of the law: as the generalized rules for society they reflect the 'common interest' or 'normality' that can legitimately be coercively enforced, even by violence. Here we find the Machiavellian duality of consensus and coercion in successful ruling strategies.

In addition to these structurating effects as captured in the three levels of the MLP, Gill also introduces a concept capturing self-governing effects of worldviews, norms and common

206 Stephen Gill, 2002, Power and Resistance in the New World Order, Palgrave, pp. 116-138.

In the end, each group or niche is the result of individuals making the choice to come together, and it will be individual people formulating the new principles for pioneering activities.

sense within individuals. *Disciplinary neoliberalism* refers to the definition of discipline used by sociologist Max Weber. It holds that classes, status groups, political parties and the like are social phenomena expressing the distribution of power in a society. They discipline those who wish to be part of these communities or networks: "What is decisive for discipline is that obedience of a plurality of men is rationally uniform."[207] In effect this means that everyone seeking to fit in with a market society develops rationales, habits and social practices that allow for him or her to lead a successful life under the hegemonic paradigm or narrative and the organizational logics or new constitutionalism patterns that have been set up.

Gramsci himself therefore urged not to restrict the idea of coercive rule to official laws but to understand how the 'private' context equally defines codes of conduct and shapes the limits of possible deviance as long as 'fitting in' is still the motivation:

> *"Question of the 'Law': this concept will have to be extended to include those activities which are at present classified as 'legally neutral', and which belong to the domain of civil society; the latter operates without 'sanctions' or compulsory 'obligations', but nevertheless exerts a collective pressure and obtains objective results in the form of an evolution of customs, ways of thinking and acting, morality, etc."[208]*

Thus, Gill's neo-Gramscian concepts substantiate the general MLP view on societies with a political economy interpretation of current path dependencies as having mental as well as legal quality. They summarize manifestations of the neoclassical paradigm and mindset at each of the three levels distinguished by Geels and Kemp. This shows how the hegemonic narrative of making sense of why things are the way they are translates into structural power potentials in change or resistance strategies: those players able to present their approach or proposals and principles as relevant or in line with the widely accepted story and goal definitions for development are likely to find support.

207　Max Weber, 1963, quoted by Gill 2002, p.130.

208　Gramsci, 1971, p.242.

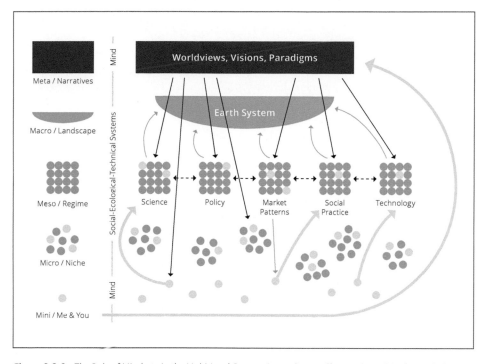

Figure 3.3.6 – *The Role of Mindsets in the Multi-Level-Perspective on System Change. From Göpel, M., 2016. The Great Mindshift. Springer. P. 47.* http://www.springer.com/de/book/9783319437651.

To capture this mediating role of mindsets or paradigms in societal transformations one can expand the MLP graph. The landscape level can be divided so that qualitatively very different aspects are separated. At the *macro level* I locate more physical-structural parameters on which humans have only very indirect and slow influence, like earth system processes. At a *meta level* I distinguish the role of worldviews or mindsets that although very resistant to change are directly constructed by humans and therefore also directly changeable, even in the short term. They permeate the social regime and niche activities that I group under the term 'social technologies', for which others may use the term 'institutions'. In addition, I added the micro level to highlight the role of individual sense-making – which can have reifying effects, as Gill pointed out, but also is the ultimate source of deviation from the hegemonic narrative or paradigm. In the end, each group or niche is the result of individuals making the choice to come together, and it will be individual people formulating the new principles for pioneering activities mentioned by Leadbeater. By connecting with complex system innovation approaches like that of Donella Meadows I will argue that changes at the rather intangible meta level translate into the potential to become objective future reality: in the end all human-created structures are a materialization of ideas.

3.3.5. Mind Shifts as High Leverage Points in Sustainability Transitions

Systems - like the ones grouped onto the different levels of the MLP, or what the MLP as a whole describes - can be many things: including a society, a family, a corporation, a university, a forest, or an economy. A system can be defined as, "[A] set of things – people, cells, molecules, or whatever – interconnected in such a way that they produce their own pattern of behavior over time."[209] A living system like a desert, a forest, an ocean, and also a city, a university, a business, or a society can be viewed as a boundary-maintaining entity and its behaviour analysed by distinguishing three different elements or components: parts, connections and purpose. Parts need not be material in nature. They can also, for example, be people, stored information, knowledge or virtual money. Accumulations of parts, material or immaterial, are viewed as *stocks* of resources that can be drawn on as the system functions. The types of interconnections or feedback loops among parts are called *flows*: these can be energy, material or information. They determine changes in each stock's quality or quantity, depending on their feedback loops. The system's *purpose* determines what it is organized to achieve (e.g., survival, photosynthesis, winning a game, providing good education, profit--making). Understanding a system's purpose is therefore essential when seeking to understand its pattern of behavior over time.

In human-influenced systems, the purpose is of course closely tied together with the sense-making of actors and the narratives according to which stocks and flows are sought to be organized. To imagine this effect it is helpful to turn to Donella Meadows' list of leverage points for system change. It ranks possible intervention points according to their transformative potential and likelihood of actually influencing them.

PLACES TO INTERVENE IN A SYSTEM (IN INCREASING ORDER OF EFFECTIVENESS)

12. Constants, parameters, numbers (such as subsidies, taxes, standards)
11. The sizes of buffers and other stabilizing stocks, relative to their flows
10. The structure of material stocks and flows (such as transport networks, population age structures)
9. The lengths of delays, relative to the rate of system change
8. The strength of negative feedback loops, relative to the deviations they correct
7. The gain around driving positive feedback loops
6. The structure of information flows (who does and does not have access to what kinds of information)
5. The rules of the system (such as incentives, punishments, constraints)
4. The power to add, change, evolve, or self-organize system structure
3. The goals of the system
2. The mindset or paradigm out of which the system – its goals, structure/rules, delays, parameters – arises
1. The power to transcend paradigms

---

209 Donella Meadows, 2009, p. 2

This list is designed for social system change and change. Adjustments at lower levels can alleviate immediate pressures but will usually not break path dependencies perpetuating the system's behavior and development dynamic. Impacts on the behavior and purpose of the system arise only if changing these lower parameters triggers leverage points higher in the list. These leverage points are embedded into many path dependencies or feedback loops: material infrastructure, social processes and institutions, and individual understanding of what is at stake in the given context. This makes them more difficult to change but also means that changing them successfully will bring more lasting change to the entire system.[210]

For example, if a government increases the minimum wage by ten percent, does this mean the former rate was not enough to pay for existence – or is it the measure expressing a new goal that income differences between people working in the same organization should be reduced? Is it simply a measure to reduce poverty statistics and entry into welfare programs, or is it a move to alleviate inequality in a society? The latter would be a new rule indicating a goal-shift as to how much inequality we accept in sustainable societies - rather than alleviating a symptom of systemic reinforcement of inequality.

Changing the third highest leverage point, the system goal, therefore means mobilizing many lower leverage points. Finally, the top two encapsulate the transformative potential of mind or paradigm shifts: "[T]he mindset or paradigm out of which the system – its goals, structure, rules, delays, parameters - arises; and the power to transcend paradigms."[211] They provide the reference framework for what seems adequate, rational, desirable and possible.

Linking these analytical categories to the challenge of sustainable development today, we see that the most important stated goal of our societies is economic growth. It prevails even though the type of growth structure we have today has many social and environmental costs. Yet, in order to tackle those problems, societies keep on pushing for more growth in order to pay the costs of the environmental and social damage. Given the ongoing neoclassical paradigm and its blind spots this path can be promoted in ignorance to the fact that most of the damage is happening because of *the type of growth* to which this development idea or paradigm aspires.

Meadows points out that this phenomenon is typical: people have a feel for where leverage points sit but often tend to push them in the wrong direction. To address negative outcomes and emerging structural frictions, the collective will suggests pushing for yet more growth of this kind instead of considering that less growth could actually reduce this damage.[212] A prominent example of this paradox is the Brundtland report which gave the original definition of sustainable development: it suggests fostering three percent growth in GDP worldwide

---

210 Donella Meadows, 1999, Leverage Points. Places to Intervene in a System, p.3. *http://donellameadows.org/archives/leverage-points-places-to-intervene-in-a-system/*

211 *Ibid*, p. 3.

212 *Ibid*, p. 1.

because otherwise rich countries will not invest in poor countries and rich people will not accept redistributive measures unless they derive from further gains. Imagination of possible future developments is limited by the image and logic of a capitalist market society connecting up to 10 billion selfish maximizers of personal utility.

> *Changing the deeply embedded neoclassical paradigm, its political and economic institutions, distribution patterns, life styles and identity-shaping discipline therefore amounts to no less than repurposing human development.*

Changing the deeply embedded neoclassical paradigm, its political and economic institutions, distribution patterns, lifestyles and identity-shaping discipline therefore amounts to no less than *repurposing* human development. This is a huge task, will involve substantial conflict, and will take time. But even very deep social structures ultimately depend on humans reproducing them. So just as paradigms and hegemonic mindsets have a hampering effect on alternative proposals, cultivating future capacities to challenge and change them also has tremendous emancipatory power. Gramsci called this the progressive self-consciousness:

> *"The awareness of self is re-constituted through an appreciation of prevailing thought-patterns and the nature and distribution of life-chances. Hence the moment of self-awareness leads to a more complex and coherent understanding of the social world and is a form of historical change."*[213]

So, while changing paradigms is not an easy task, applying reflexive futures literacy practice is immediately open to everyone – as the direct arrow from mini to meta level in Figure 3.3.6 expresses. Donella Meadows cites Thomas Kuhn when outlining this path of engagement:

> *"In a nutshell, you keep pointing to the anomalies and failures of the old paradigm, you keep speaking louder and with assurance from the new paradigm, you insert people with the new paradigm in places of public visibility and power. You don't waste time with reactionaries; rather you work with active change agents and with the vast middle ground of people who are open-minded".*[214]

A lot of this is happening today, as the practical examples in the next section show.

---

213 Here quoted by Stephen Gill, 2002, p. 31.
214 Donella Meadows, 1999, p. 18.

3.3.6. Living a New Economic Paradigm in Practice

Across the world many different initiatives and movements for alternative ways of organizing human relationships and human interactions with nature are in place and emerging. We see a wave of small-scale *repurposing* experiments with alternative systems of production and consumption. They are of different sizes and shapes and carry different names, but their commonalities in paradigm are striking. None of them sticks with the story that actors are selfish and insatiable independent units or that market prices and efficient competition are the only goal for successful development processes. All of them track connections between social processes and those of the natural ecosystems around them. They seek to understand how the wider system context influences actor decision-making and institutional development trajectories. The following sub-sections review four exemplary movements that are rapidly growing. It is a very Eurocentric snap-shot of initiatives that have achieved mainstream attention. Further research could seek to map many more initiatives around the world and document paradigmatic similarities or differences.

The Economy for the Common Good

The mindset of this movement does not extrapolate from the description of 'how humans are' but starts from a societal or system view. It locates the challenge of successful sustainability solutions for thinking and aspiring individuals in the balance between community responsibility and individual freedom. Neither can work without the other: individuals need cooperation to flourish and build wealth and the community needs creative deviators in order to diversify and adapt.

Christian Felber, a lead author in this movement, therefore emphasizes the need to reconnect private entrepreneurship with the overall binding goal of the common good. The latter can only be defined in democratic political processes and the former shows how a particular way of running a business can deliver on it. According to Felber, current economic rules encourage egoism, greed and striving for power. Despite what the 'com' in 'competition rules' would suggest (Latin for 'together' or 'we'), they ensure that winners basically take all and render even hostile takeovers of entirely healthy businesses as a legitimate outcome as long as purchasing power can push it through. This incentivizes relationships of 'contrapetition' in which asocial and anti-social behavior pay off, attacking units are strengthened for the next battle and power and wealth are increasingly concentrated.[215]

> *Economy of the Common Goods emphasizes the need to reconnect private entrepreneurship with the overall binding goal of the common good.*

215 Christian Felber, Gemeinwohlökonomie. Das Wirtschafsmodell der Zukunft, Deuticke, 2010.

The movement's website *www.gemeinwohloekonomie.org* proposes 20 principles for how an alternative type of economy could be put into practice. Rather than fixed rules, they are seen as inspirations for reflection and dialogue on the values, norms and practices that status quo institutions and regime solutions nurture or even prescribe. Principle one expresses the overall mission purpose:

> *"The same collectively shared values that contribute to fulfilling interpersonal relationships are the basis for the Economy for the Common Good: confidence building, cooperation, appreciation, democracy, solidarity. Scientific research proves that fulfilling interpersonal relationships constitute a key factor to happiness and motivation."*[216]

Following from this the foreseen 'more intelligent rules of the game' should lead away from contrapetition towards cooperation, from personal profit to common good, and from market control to democratic decision making. In an Economy for the Common Good, business performance measurements therefore go beyond the internalization of environmental harm into market prices: "Economic success will no longer be measured with (monetary)

> *The entire idea is to make selfish and ruthless behavior the more difficult and costly option rather than the easy and profitable one.*

exchange value indicators, but with (non-monetary) use value indicators." As a consequence, similar indicators for business and economic performance on a societal level can align bottom-up and top-down initiatives towards the new societal purpose:

> *"On the macroeconomic level (national economy) the Gross Domestic Product (GDP) will be replaced – as an indicator of success – by the Common Good Product. On the microeconomic level (company) the financial balance sheet will be replaced by the Common Good Balance Sheet (CGBS). The CGBS becomes the main balance sheet of all companies. The more companies act and organize themselves along social, ecological and democratic lines, the more solidarity they display, the better will be the results of their Common Good Balance Sheet. The better the CGBS results of the companies within a national economy, the higher its Common Good Product."*[217]

The rules and incentive structures of Common Good Economy are not to be confused with a socialist planning state that Felber himself diagnoses to have suffocated individual freedom. The entire idea is to make selfish and ruthless behavior the more difficult and costly option

216 See the website *http://www.gemeinwohl-oekonomie.org/en/content/20-principles-guiding-economy-common-good*
217 Ibid

rather than the easy and profitable one, e.g. creation of social and environmental costs would now incur a competitive disadvantage rather than advantage. A research area full of relevant ideas on how to redesign institutions, infrastructures and social settings so that they support decision-making in line with particular goals is Behavioral Economics. Here the term 'nudging' refers to non-coercive structural factors that enable rather than hinder sustainable behavior in any given situation.

The Transition Movement

The Transition Movement finds its common denominator in engaging people in collective change processes in communities of place. Originating in the UK, it has spread to at least 43 countries worldwide[218] and makes 'reflexive relocalization' its core stance. The term reflexive is important because it highlights that the change processes are driven by communicative engagement among members and not imposed by rule changes and control of compliance. The general paradigm pictures communities as social-ecological systems embedded in wider environmental systems; the goal is to improve the resilience of the community in light of growing megatrends like climate change, rising energy prices and economic crises.[219]

In the first Transition Handbook, Rob Hopkins, founder of this movement, describes resilient sustainable communities as those that are structured along three principles: *diversity* of life-supporting solutions or livelihoods, *modular structuration* with buffers to the outer systems that increase self-reliance possibilities, and *tight feedback loops* that bring the results of actions closer to those responsible for them.[220] This of course is easiest done at the local level where physical proximity facilitates compliance with these design principles

Once again, overarching system dynamics determine which production processes are promising and the assumption is that learning actors rationally adapt their solutions accordingly. Rational in this context, however, means with reason and a lot of discussion rather than an automated response to cost-benefit stimuli. Part of this reasoning involves assessing foundational ideas around what humans need and want and questioning whether efficiency gains are always good. An explicit part of increasing self-reliance and resilience, for example, means turning away from massive economies of scale that are only possible under systems with very high divisions of labor and concentration of production. Less mass production and a focus on non-consumption strategies for wellbeing are also central elements and pursued by linking, "[S]atisfaction and happiness to other less tangible things like community, meaningful work, skills and friendship."[221]

218 *http://www.transitionnetwork.org/initiatives/.* Accessed October 27th 2014.

219 Rob Hopkins, The Transition Handbook. From oil dependency to local resilience, 2008, p. 10.

220 *Ibid,* pp. 55-56.

221 Rob Hopkins, 2012, Resilience Thinking, in: Bollier and Helfrich, *The Wealth of the Commons*, The Commons Strategy Group, pp. 20-21.

The vision behind Transition Towns or communities is one of a resilient world built on the promotion of trust, well-developed social networks, and adaptable groups working well together. Research strands providing evidence that this will create more happiness than neoclassical ideas of endless individual competition for more consumption include positive psychology, wellbeing studies and neuroscience. Many more principles of the Transition movement contradict the notions of neoclassical models: actors are explicitly requested to change their way of thinking and being in the world and to share instead of compete. Production and cooperation processes are intentionally designed to be less efficient and centralized in order to increase resilience and co-creation.

> The vision behind Transition Towns or communities is one of a resilient world built on the promotion of trust, well developed social networks, and adaptable groups working well together.

The main mission is summarized as follows: "To inspire, encourage, connect, support and train communities as they adopt and adapt the transition model on their journey to urgently rebuild resilience and drastically reduce CO_2 emissions."[222] Once again, the economic system is viewed as a subsystem of socio-ecological systems that should serve this higher purpose and fundamentally change if necessary. The emphasis Transition hence places on collaborative economic solutions embedded within the social and ecological realities of place lead some analysts to locate it within a broader global movement for the defense and creation of commons.[223]

Commoning

According to the Commoning movement, a fundamental change necessary for resilience and sustainable prosperity is the dethroning of private property. Commoning solutions envision and enact non-commodified relationships in which joint responsibility for the maintenance of the overall system is an integral part. At the center of these governance approaches lies an ideal of property that treats most of what exists today as the common heritage of humankind to which each person is equally entitled. This implies that each generation should not use more than future generations will need to enjoy similar levels of wealth. Jointly produced value is conceptualized as a common good outcome rather than divided into individual shares of the market returns in line with the particular 'value' that each contributor brought to the process. Thus, commons imply both responsibilities and benefits: alongside being co-stewards of

222 Rob Hopkins and Peter Lipman, Who we are and what we do, document on the Transition Network website, for download at http://www.transitionnetwork.org/sites/www.transitionnetwork.org/files/WhoWeAreAndWhatWeDo-lowres.pdf

223 Justin Kenrick, 2012, The Climate and the Commons. In B. Davey (ed.), Sharing for Survival. Feasta. http://www.sharingforsurvival.org/index.php/chapter-2-the-climate-and-the-commons/: Henfrey & Kenrick, this volume.

Figure 3.3.7 – Edible Cities Conference, Witzenhausen, 2013. Credit: Gesa Maschkowski.

what earth and ancestors have provided, everyone is conceived to be a co-proprietor of the wealth created.

The book *The Wealth of the Commons: A World Beyond Market and State* comprises 73 essays from thinkers and practitioners in the field. The commonalities binding this rapidly growing community are described as 'an overarching worldview' along with a set of social attitudes and commitments and a political philosophy or even spiritual disposition that guides an experimental pathway for strategic change.[224] The introduction of the book highlights statistics that show how much 'overwealth' (Überfluss) there is in the world and that it is not scarcity but unsound patterns of production, distribution and consumption that create the unsustainable outcomes of today. Thus, it is also not simply a question of better technologies but of better institutions with their psychological, socio-cultural and institutional path dependencies.

While there does not exist one unitary definition of the commons or commoning, one website central to the movement (*http://onthecommons.org*) summarizes the gist of this paradigm. The core principles characterizing all commons initiatives are:

224 David Bollier and Silke Helfrich, 2012, The Wealth of the Commons, The Commons Strategy Group, pp. xii-xiii.

> *equity* – everyone has a fair share of our commons to expand opportunities for all;
> *sustainability* – the common wealth must be cared for so that it can sustain all living beings, including future generations;
> *interdependence* – cooperation and connection in communities, around the world and with the living planet is essential for the future.

The characteristics of community life in line with the commoning vision are described as:

> *shared governance* in the most participatory form;
> *deepened responsibility* for the restoration and care of the common inheritage;
> *belonging* as a general outlook on ownership and organization;
> *co-creating* as a form of engagement and sharing that highlights the abundance of skills and solutions rather than scarcity.

Commoning approaches therefore distinctly break with the organizational logic of markets and declare the profit motive and individualistic competition processes to be core drivers of unsustainable solutions.

3.3.7. Conclusion

Before an individual decides to act, he or she requires a story or mindset to make sense of what life is all about and what is at stake in the given situation. Acting rationally from a reflexive science view therefore means first and foremost to act in congruence with one's worldview, and with one's interpretation of the social logics or 'rules of the game' and if those can or should be changed. Research designs treating mindsets or paradigms as core variables therefore seek to show how the same situation or possible future developments are viewed very differently depending on the chosen lens. The goal of this paper was to connect a critical political economy approach in reflexive science, namely Gramscian hegemony theory, with currently widely discussed concepts in transition theory and the notion of futures literacy.

By discussing exemplary manifestations of the neoclassical paradigm with reference to the Multi-Level-Perspective on societal change it showed how ideas and their materialization in concrete norms, practices, rules, laws, material infrastructures and physical technologies create a framework of action that influences how human needs develop as much as what seem adequate or possible solutions for commonly defined goals. It therefore engages with the research challenge that the 2013 World Social Science Report also formulated:

> *"Critical to a social-ecological systems perspective is the role of humans as reflexive and creative agents of deliberate change. Understanding how values, attitudes, worldviews, beliefs and visions of the future influence system structures and processes is crucial".*[225]

225 World Social Science Report 2013. Changing Environments, UNESCO and ICCS, Summary, p. 7, online readable at *http://www.oecd-ilibrary.org/social-issues-migration-health/ world-social-science-report-2013_9789264203419-en*

Figure 3.3.8 – Shoe bed at Allmende-Kontor Community Garden. Credit: Gesa Maschkowski.

Applying futures literacy like the examples of practice reviewed in the previous section allows envisioning and creation of institutions, processes, technologies and business models that are sustainable by design rather than relying on cleaning up after the event. It also empowers actors to identify and speak up against the co-optation of new ideas, frames and narratives into the old paradigm so that their transformative potential is contained.

The paper showed that neoclassical worldviews and models are full of blind spots regarding the origins and qualities of human needs as well as natural processes providing the resources needed to satisfy them. It also provided some first ideas as to where their nevertheless ongoing reification is located, from overarching infrastructures to individual identity formation. As philosopher Richard David Precht points out:

> *"Strict and tough calculation of utility, ruthlessness and greed are not man's main driving forces, but the result of targeted breeding. One could call this process 'the origin of egoism by capitalist selection,' following Charles Darwin's famous principal work."*[226]

My final conclusion therefore holds that the ultimate drivers of societal change are located within each one of us. In comparison to the magnitude of the challenges that earth scientists

226 Richard Precht, here cited by Habermann 2012, We Are Not Born as Egoists, in Bollier/Helfrich 2012, p.15

and poverty statistics describe this may sound disproportionate. But each act of doing things differently, each questioning of purpose or reasons, leaves a dent in the former framework of action and its reifying impacts. Psychology, sociology, neurosciences show that shifting mindsets implies not only a change in thinking but a change in being, feeling, engaging, relating and acting in the world. They are at the root of what we can imagine as possible sustainable futures and adequate social as well as physical technologies and governing systems to host them. Various fast-growing pioneer movements for new sustainability solutions are an expression of this. They all carry clear principles and imaginations of system designs whose purpose is a different one than economic growth and market forces.

The fascinating work of the next years and decades in research and practice will be to keep on working out the new paradigm or storylines emerging from this movement and see how they may shape into a new collective will with a compelling Gramscian social myth. The latter needs more conscious storytelling and strategic coalition building among pioneering initiatives or change makers highlighting these niche practices to argue for change at regime levels. After the first superficial comparison of paradigms behind Common Good Economy, Transition Movement and Commoning I am less pessimistic than sociologist Harald Welzer that this is possible:

> *"For the time being, the transformation necessary today lacks guiding principles of the kind that early industrialized societies had in terms of progress, freedom, prosperity and growth. It will not be possible to establish new mental infrastructures without guiding ideas, yet if they do not dovetail almost naturally into day-to-day lives and lifestyles, visions of the self and frames of reference for the future, they will remain just that – ideas."*[227]

Instead, I argue that we do not need to reinvent principles but much rather reclaim the meaning of what deeply anchored human values are connected with. Among the pioneers we find overlapping ideas for this: a holistic understanding of *prosperity* beyond consumption needs that guides equitable and balanced *progress* of the whole socio-ecological system to improve human security - *freedom from fear* to fall behind or to be enmeshed in conflicts over resources and *freedom from want* that marketing and advertising constantly create. The examples already show that this leads to an unprecedented *growth* in the *creativity* of strategies for satisfying non-material needs and *conviviality* in the processes for enacting them. As a result we can add another benefit for future human development: improved individual, communal and societal *resilience* in a world whose transformation - towards sustainable development or in any other direction - will present us with a rocky ride.

227 Harald Welzer, 2012, Mental Infrastructures, essay published by the Boell Foundation, Germany, p. 32.

RESILIENCE IN PRACTICE

4

4.0. The Essence Of A Resilience Approach To Management And Development

BRIAN WALKER

1. Whatever the management or planning/development issue, put it into a whole-system context by developing a conceptual systems model in which the issue is embedded – a linked social (including governance) and environmental model.

2. Especially, consider and approach the issue at multiple scales. Probably the most common mistake made in planning/management is to restrict the analysis to the scale at which the issue is expressed. You cannot resolve the problem by focussing on only one scale. It's essential to consider the scales above and within, and to identify the main connections/influences across those scales.

3. Using the evolving conceptual model, ask the question, "The resilience of what, to what?" to identify what is of most value in and from the system, and the major threats and disturbances the system faces. Use this to start the process of identifying where the system is weak in regard to its coping capacity, or general resilience (again, in terms of both social and environmental variables).

4. What are the most significant, critical limits/thresholds you need to avoid crossing, and what processes are driving the system towards them? This identifies the crucial controlling variables in the system.

5. To manage these dynamics, where is it necessary to: i) use adaptive management to build the resilience of the system; and

Figure 4.0.1 – Brian Walker. Credit: Gesa Maschkowski.

ii) reduce resilience of the existing system to enable transformational change. The latter is needed when the system (or parts of it) has shifted irreversibly into an undesired state, or is about to do so.

6. Consider (again) cross-scale effects and trade-offs. To avoid crossing a threshold at one scale it may require transformational change at another scale.

Figure 4.0.2 – *Multilevel Rocks. Credit: Gesa Maschkowski.*

4.1. Open Space Write-up: Resilience, Community Action and Social Transformation

TRANSITION RESEARCH NETWORK; WRITE-UP BY THOMAS HENFREY

This chapter reports on a parallel session at the Resilience 2014 conference on Tuesday May 6th 2014 in Montpellier, France. The session aimed to examine relationships and interactions between practitioners' situated and experiential knowledge and the more abstract and theoretical understandings of resilience researchers in a range of disciplines.

Figure 4.1.1 – *Open Space Session.*

Figure 4.1.2 – *Mutual Learning from Transition and Resilience Research. Credit: Gesa Maschkowski.*

The session was convened by ECOLISE, the European Network of Community-Led Initiatives on Sustainability and Climate Change, whose member organisations have up to 50 years of practical experience in promoting, creating and living within sustainable communities. It was organised by practitioners and action researchers within grassroots movements for resilience: Tom Henfrey, coordinator of the Transition Research Network and Researcher at the Schumacher Institute for Sustainable Systems in Bristol, UK; Juan del Rio of the Spanish national Transition hub; Lorenzo Chelleri of the Barcelona Autonomous University; Gesa Maschkowski, a PhD researcher at Bonn University and member of the German Transition Network; and Glen Kuecker, history professor at de Pauw University in Indiana and active supporter of indigenous struggles in Latin America.

About 30 researchers and practitioners attended. Following brief presentations from Maschkowski, Chelleri/del Rio and Kuecker (respectively the basis of Chapters 3.1, 2.2 and 4.3 in this collection), it adopted an interactive discussion format based on Open Space technology. Within the general theme of resilience and community action, all present were invited to propose discussion topics. The group agreed on four questions:

1. What can transition movements and resilience research learn from each other?

2. Measurement versus process – what are the ways forward?
This is a reaction to the emphasis on large-scale quantified measurement in some of the other sessions, and might link to the next question on resistant institutions.

3. Working with and within Resistant Institutions.

How to bring ideas of 'the salutogenetic way' into institutions and planning? How to release resilience or transition understanding into resistant institutions?

4. Drivers for Involvement in Transition

Going beyond the salutogenetic approach, which is very important but can't explain everything: What are the main drivers of people's involvement in Transition initiatives?

How can the resilience assessment methodology employ the sense of coherence mentioned?

To what extent are motivations of instigators the same as the drivers of people involved in the broader movement?

People formed breakout groups around each of these questions. Everyone present initially joined the table whose question most interested them. People were free to move to another table at any point – though in fact in this session nobody did. After a 20-minute discussion, one participant from each breakout fed key points back to the group as a whole. Breakout sessions were recorded: sections 4.1.1 to 4.1.3 are based on transcripts of these recordings; 4.1.4 (whose recording was inaudible) has been reconstructed from notes taken during the session.

4.1.1. What Can Transition Movements and Resilience Research Learn from Each Other?

Summary:

1. Having clearly stated goals and agendas means Transition practice tends to employ specific and normative concepts of resilience. Researchers tend to use resilience in ways that are more general and conceptual but that may obscure hidden norms and values.

2. Transformative agendas in Transition imply a need to question many established norms and values. Some resilience research does the same, but much employs a more conservative discourse of 'bouncing back'.

3. Transition is a social experiment based on learning by doing whereas research tends to involve learning by observation [and analysis].

4. Researchers and the Transition movement are already learning from each other in important ways.

Revealing and challenging hidden politics of meaning behind different uses of resilience. Following a point that came up in a session on urban transformations the previous day: there's a tendency to use resilience as if it was a normative concept that implies positive values, and that isn't necessarily true. In Transition there is a clear and explicit set of objectives relating to sustainability and wellbeing in the face of declining net availability of energy, climate change, and financial instability. This implies a very specific type of resilience and a clear trajectory towards it involving things like self-reliance, building community through collective action, and promoting equality. Resilience thinking in a technical sense is a framework that can be applied towards many different ends that may reflect very different sets of values and desired objectives.

So if researchers and resilience practitioners want to talk about resilience, they have to understand they are talking about different things. These differences of definition can support really instructive interchanges. The bigger picture provided by researchers helps us understand that Transition and other forms of community action aren't about building resilience in neoliberal structures as some of the politically dominant discourses on resilience seek to do. Transition gives researchers a clear picture of what a resilience-building agenda based on explicit environmental and social values looks like, and forces researchers to take account of the hidden values that might influence the conduct and application of their research.

Understanding Interactions Across Scales and other Key Differences

One of the useful things about technical concepts of resilience is that they involve *Panarchies:* nested systems at different spatial scales, and interactions in both directions across these scales. So it takes into account influences that local scales can exert at higher levels, and how broader scales influence local action. Transition initiatives usually emphasise practical work at local scales: it doesn't of course ignore larger scales, but the local focus tends to be more prominent. Research places greater emphasis on theory and does not privilege any particular scale. This can perhaps help practice to remain attentive to its cross-scale dimensions.

> *Researchers tend to learn by watching those who are learning by doing and that perhaps gives them a different perspective. But all are part of a great societal experiment in which resilience is the key guiding concept, and in which all the different learning processes involved follow the adaptive cycle.*

Following on from the previous point, different actors, whether at the same and different scales, will have different ideas about resilience and agendas for achieving it that reflect differences in their outlooks and interests. Bottom-up and top-down action for resilience are often motivated by very different goals, or there's scope for great conflict between community-based action and municipal initiatives. Resilience thinking provides a way to understand those cross-scale actions better, and understand trade-offs between the interests of different actors and at different scales, and potential conflicts and synergies that arise.

This also relates to relationships between resilience and transformation. Many of the common definitions of resilience are so much about bouncing back, maintaining structures and identity of the system, whether that's the community or something else, but resilience is also about the capacity to change and deal with uncertain outcomes. So one of the questions a Transition initiative is asking is how to maintain the core structures and functions of a community or system and at the same time maintain a capacity to change and go forward without knowing the future. In looking for practical answers to this question, it identifies needs for change in parts of the system which are no longer useful, viable or desirable, so in practice Transition implies a need for transformation. This seems different from how many researchers use the term transition in a technical sense, which tends to mean increases in efficiency and effectiveness, at times conservative but including addition of new structures and functions. Transformation is a deeper concept that involves questioning existing norms and ways of doing things the way the Transition movement is doing.

Another question this raises is who is to decide what change is desirable or necessary. So you have to think about who leads, who makes the decisions and in whose interests. This includes who becomes vulnerable – a resilient system will include areas whose vulnerability is beneficial to the system as a whole, so how do you make that trade-off?

Learning by Doing versus Learning through Observation and Analysis
Transition initiatives are community-scale attempts to address the local manifestations of

Figure 4.1.3 – *Measurement vs Process. Credit: Gesa Maschkowski.*

global problems. So as we've already said, they have clear objectives and can frame a specific idea of resilience against those objectives. This of course implies a second set of questions about how to achieve that, and the solutions are very different in different places. A resilient food system for a major city in which relatively few people have access to land for cultivation will look very different from in a rural community. When Transition groups identify potential solutions, this allows them to offer suggestions to decision-makers. These suggestions can also be very useful for resilience researchers.

Research and practice can be closely interlinked. There is a research branch to the practical work, as people are learning by doing. Researchers tend to learn by watching those who are learning by doing and that perhaps gives them a different perspective. But all are part of a great societal experiment in which resilience is the key guiding concept, and in which all the different learning processes involved follow the adaptive cycle. Transition can be viewed as a self-conscious attempt to create an evolved form of resilience, based on learning from present challenges and the lessons of the past. It is inherently progressive – present day challenges differ from any we've faced before, and we won't solve them just by looking back. Return to a previous state would anyway not be possible or desirable because society has already transitioned to different ways of doing things, these ways have become culturally embedded, and people don't want to return to how things were.

Resilience is a framework that can help us understand how this Transition process might look, but it can't specify the details of solutions. So research and practice are completely integrated, inseparable in fact. More broadly, we could think of our current resilience as comprising the reservoir of social capital, ecological capital and knowledge – available to use at the moment and on which Transition needs to draw: resilience thinking is an important part of this.

4.1.2. Measurement versus Process

Summary:
1. Measurement can be a problem because it creates a fixed point, whereas resilience is about dealing with uncertainty: so if we try to measure resilience against something fixed we may not be measuring what we need. So the buzz around resilience may be a drive towards simplification that closes down possibilities rather than opening them up. This touches upon the more fundamental question about ways forward: the need to change things that are currently not sustainable

2. Questions of power and process are important – whether people are involved in measurement and assessment processes, and whether the work of researchers in any way contributes to the people they are working with. That's definitely true in Transition as a co-owned process – so it's more than just information-gathering as things are definitely happening after researchers leave.

3. Problems of scale – measurement and process at micro versus macro level, global versus local, scaling up and how to affect change at broader scales and over extended timescales.

Opening Up and Closing Down

There is a potential clash between attempts to measure resilience, especially in highly quantified ways, and the understanding of resilience (and resilience building) as ongoing dynamic
processes that might be difficult to capture in a structured, scientific way.

Attempts at measurement and quantification can reduce resilience to some measurable, tangible attribute of the system in ways that are at odds with more process-oriented approaches.
This is particularly true when the purpose of measurement is to improve the ability to control
the system. When something is made measurable, it is made legible – visible, knowable - in
ways that foreclose possibilities. In that process measuring may limit the system in important
ways: defining what resilience is in a certain way to the exclusion of others, specifying what is
to be measured and hence what is not to be measured, and limiting the way the system can
develop or evolve in the future.

Benchmarking in order to measure and assess outcomes is also important here. When you
benchmark you're specifying the goals and objectives, effectively setting future targets
against which to measure progress towards resilience. Does that then exclude or deny possibilities for resilience in a way that contradicts the fluidity of the concept? The definition of
resilience used is crucial here. If it refers to emergent properties of a complex system, like its
regenerative capacity, does a defining benchmark proposition that makes things legible limit
possibilities of future emergent properties? If that sets the system on a trajectory that denies
certain possibilities, it limits the future options – and so reduces resilience. If resilience is the
capacity to deal with situations that are unpredictable or can not be anticipated, that makes it
really hard to measure anything meaningfully against a fixed reference point.

Brian Walker suggested in the morning plenary session that it makes no sense to define
resilience as a goal that you want to reach. It's a far more open process, and that makes
it extremely difficult to define criteria. It's also context-dependent, so changes in context
over time make meaningful measurement and comparison far more difficult, perhaps even
impossible. If in ten years you measure the same criteria but the context has changed, those
criteria will no longer be relevant and what you measure will not have the same implications.
So measurement may not be such a good idea. It's also important to address issues around
timescales and the need for a long-term approach.

Clarification and Appropriate Methodology

On the other hand, attempts to define and measure resilience force a clarification of meaning.
This can oppose the trend to use resilience as a buzzword whose meaning is inconsistent.
For that reason it's important to operationalise certain concepts, but that needn't imply
quantifying them in rigid ways. One alternative could be to use relative measures. There
are some interesting concepts from network governance that could be used to assess
community resilience. Innovative capacity is one: if you could measure this and meaningfully
compare it across different communities, you might be able to say that one has a higher
innovative capacity than another, and this is likely to mean it is more resilient. That's one of

many approaches we could potentially adapt. The question is how we interpret and apply our measures – not as absolute numbers, more as a relative moment in time. If we allow things to remain too fuzzy, there's risk like with sustainability that the concept loses all power.

If we explore ideas of qualitative measurement, this leads us into some interesting possibilities for constructing narratives: for example about the way you live or want to live, and about the possibilities that you see yourself being able to change. That makes comparison more difficult, but it's far more likely to be meaningful. There's a discussion about whose story gets told – and what becomes the prominent story – but in some ways those discussions are the whole point. I see the process question as basically a power question: who decides, what are the methods?

Innovative capacity is one interesting concept that could be used to assess community resilience: if you could measure this and meaningfully compare it across different communities, you might be able to say that one has a higher innovative capacity than another, and this is likely to mean it is more resilient.

We can avoid potential clashes between measurement and process, by being very careful about what we measure and how those measurements are interpreted. In some ways we need to make things more concrete and quantified measures can have a useful part to play in this. It is important to be very clear about what that can and can not tell us, to be careful about how the information is used and communicate clearly exactly what is being measured and why.

Scale and Timescale

Issues of scale can illustrate how inappropriate measurement – or interpretation of measurement – can narrow the analysis in a way that limits possibilities. For example, if you measure at a very large scale, you lose lots of important details of context. Large scales also depersonalise people's engagement with issues as the question, "What can I, as an individual, offer to this process?" gets lost. Talking about smaller scales where people have direct experiential knowledge of their system can be part of a process that is about more than itself: it creates something that will be taken out of the room and live on. This is an important point in all research: when researchers gather information, are they just extracting that information and leaving, or is it part of a process that will improve people's capacity to build resilience in the long term? It's important to look for ways in which local processes can both yield information and create something that remains in the community once the research if finished.

If the core question behind Transition is how we work together to change the world, there's perhaps a tension or contradiction with resilience. If resilience is defined as the ability to absorb shock without the system fundamentally changing, it may hinder the change that is necessary. So resilience might become part of the way the system perpetuates itself. Especially if we reinforce that by forms of measurement that lock the system into a certain path or direction, that may reproduce the system as it is rather than opening the possibilities needed for Transition to happen. It depends how resilience is defined: some definitions include the capacity to transform when the current system is no longer viable, or to anticipate a different system, or to work at a different organisational scale – for example, managing different communities at a regional level.

4.1.3. Working With and Within Resistant Institutions

Summary:

1. Those working within incumbent institutions or seeking to collaborate with them often experience deep-seated resistances to change that contrast with the liveliness and dynamism of Transition and other social movements.

2. Confronting and overcoming these institutional inertias requires courage, stamina and strategic action.

3. Bureaucracy can be as alive as any other part of the system, and a potential locus of activity as well as inertia – depending on the level of disconnect.

4. Normative biases can arise when we employ specific constructions of reality, implicitly or overtly linked to uses and definitions of particular terms – including 'resilience'.

Radicalism versus Inertia

Institutional inertia is a feature of many organisations working at community level. This can include research institutions, charities and large NGOs. People committed to positive change and working in institutions that are unwilling to accept the prospect of major societal transformation often experience powerful tensions, and find themselves surrounded by people who either don't fully grasp the need for change or are really frightened by the prospect. It's very difficult to bridge the gap between such situations and those in movements like Transition towns in which the need for transformation in current systems is widely accepted. When a single person or a small group contradicts a deep-rooted and powerful consensus against change it's easy to be marginalised or get squeezed out. A key task of people seeking to facilitate these bridging conservations is to help grassroots actors understand power, and so learn how to engage with and influence established institutions.

Some community-level stakeholder organisations such as local governments, schools, and established environmental groups tend to use key concepts such as sustainability in static or retrograde ways. Resilience is a more dynamic and less familiar concept, so it's possible its

Figure 4.1.4 – *Group Discussion. Credit: Gesa Maschkowski.*

use could encourage more flexible thinking and dialogue. This is by no means certain: fundamental change requires challenging many established world views, and success will always be limited if these world views continue to constrain possibilities.

The Courage to Work for Change – Some Experiences

What I'm seeing in my community is that as well as just speaking about or promoting change, it's important actually to do it: to demonstrate change through transformative projects like a community currency scheme. Then if after doing all this stuff you still have the time and energy, you can start to measure and evaluate: to ask whether it has led to any social change or affected dominant mindsets. We're just getting to that point in my community, and I was sponsored to come here to learn how to put that into a bigger context: how to take it further and be more influential when confronting the expertise that speaks against it. This takes a lot of courage: as a single volunteer sitting on the municipal table that represents 20 municipalities, it's not easy to speak out and not to be squashed by the dead hand of bureaucracy and the inertia it creates.

In my short experience as a city councilor I found the best way was just to turn up every day, to be there, and to see who supports me, slowly over time. We call that a dialogue of alignment, a strategy of slowly finding alignment among different stakeholders. Even if we have some fundamental disagreements, over time we can work out where we're aligned and so find ways of collaborating for change.

The Dead Hand of Bureaucracy, Living Hand of Social Action

Building on the theme of courage: I work in a very large organisation where people can hold onto their initial ideas very tightly. If you release your idea, it can often change into something that is more meaningful to the majority but not true to your original aim. This can be true in social movements as well as institutions, and can prevent an idea from reaching its full potential. So it's important to look at the social and organisational processes through which someone's idea turns into collective action.

There's a contrast between the dead hand of bureaucracy and a mode of working that allows things to happen as opposed to one that seeks to keep things a certain way. The difficulty is mobilising in a world of bureaucracy for a movement that is premised on very different means and organisational processes. The negotiation between them could be something like a research instrument, or activity instrument, and the negotiation between them can be quite profound. Another way of expressing it is simply to call it power.

Epistemology, Knowledge and Power

If we think about the question of how to bring the idea of salutogenesis into institutions, an important part of the answer is: with caution. What Gesa talked about was something different – simply a framework for communication about what people want from an organisation – the original idea follows quite a normative notion of 'what is health', and that is also true of community health. This raises a general point of the importance of being mindful of what constructions of reality particular ideas imply, and on that basis considering whether and how we want to bring them into institutions and the likely consequences. This is relevant to the whole

> *An example is the work of Phil Cass in Ohio: Over time they boiled their enquiry down to a single question: What is health? By asking that question to many people over an extended period they changed the health system.*

of the resilience discussion: resilience itself is a normative idea, we should be cautious with that. It's an example of post-colonial understanding of power and knowledge: the hidden power relations underlying how a term is defined and used, or even when you say 'they' rather than talk about what we need and what we want.

Another aspect of health is institutional health, the health of the whole organisation. We need to define health more broadly: in terms of wellbeing in addition to physical health, and in terms of the need for healthy systems. This will help us to apply these concepts more effectively to urgent questions. Implicit in this broader definition of health is to understand the factors that lead it to improve or decline: for instance how the health of the environment affects the health of individuals, and the effects of inequality on everyone's health. An example is the work of Phil Cass in Ohio, a seven-year collaboration with a group of health providers. Over time they boiled their enquiry down to a single question: What is health? By asking that question to many people over an extended period they changed the health system.

All of these are examples of the more general challenges of stepping into a transdisciplinary space that requires you to adapt your own expertise to novel contexts and rely on the expertise of others in areas in which you may not be familiar. When ecologists and others with technical or scientific expertise have to work on engagement and partnerships, for example, it brings in questions of social change processes for which they might not have the relevant skills. It's much more difficult to work it out for yourself than bring in other people that already have those skills, but institutional inertia can often make it difficult to assemble multidisciplinary teams. This can often lead to simplified processes that simply drive certain agendas.

4.1.4. Drivers for Involvement in Transition

Motivations for involvement change over time: this is true both for individuals and for the group.

People get involved for reasons that may be different from those that motivate them to sustain their involvement. They themselves change, their situations change as a result of getting involved, and the groups and projects themselves change so that they offer different things.

So the people who get involved in the early stages may have very different motivations from those who get involved later, when the project or initiative offers very different things.

Shared norms and values are agreed in the early stages of a group's life: people feel included when they are able to take part in this, and later may be attracted when they find these norms fit their own. This is another reason that the motivations of initiators often differ from those of the people who get involved slightly later once the norms, values and activities of the group are established.

It might be useful to examine the basic needs people have (from Max-Neef, Maslow, etc.) and whether satisfaction of these needs motivates people: whether people think involvement will help satisfy these needs, whether it actually does, and the extent to which addressing these needs is taken into account in the design of group processes and activities.

The sense of inclusion and belonging is important for many people. They want to belong to a group, and co-create a 'family': an environment where they feel comfortable.

4.2. Climate, Commons and Hope: The Transition Movement in Global Perspective.
TOM HENFREY AND JUSTIN KENRICK

The climate crisis has revealed a paradox at the heart of global governance:[228] those who hold the power in the current system present as the solution to ecological destruction and social dislocation the very paradigm and power relations that are driving them. This chapter puts forward an alternative point of view and course of action.

Several widely held myths need to be challenged to bring about a just transition:
> The illusion of humanity and human economic activity as somehow separate from nature and broader ecological processes;[229]
> The assertion that competitiveness – whether among individuals, groups, nations or groups of nations – is an inevitable and/or desirable aspect of human nature; and
> The idea of a uniform, linear pattern of development through which all societies must pass in order to better their condition.[230]

Above all, the global climate crisis has shattered the idea that economic growth can be perpetuated indefinitely on a finite resource base.[231] The number of people materially benefiting from the economic and political system driven by these assumptions has grown over time, with the global financial crisis of 2008 either a brief interruption or sign of things to come.

228 Hulme, M., 2009. *Why We Disagree about Climate Change: Understanding Controversy, Inaction and Opportunity.* Cambridge University Press.

229 E.g. Wollock, J., 2001. Linguistic diversity and biodiversity: some implications for the language sciences. Pp. 248-262 in Maffi, L. (ed.) *On Biocultural Diversity.* Washington and London: Smithsonian Institution Press. Kidner, D.W., 2001. *Nature and Psyche.* Albany: SUNY Press.

230 Escobar, A. 1995. *Encountering Development: the Making and Unmaking of the Third World.* Princeton University Press, Chichester.

231 Jackson, T., 2009. *Prosperity Without Growth.* London: Earthscan. Heinberg, R., 2011. *The End of Growth.* Forest Row: Clairview.

However, this expansion has been at the expense of all who are still outside the circle of material beneficiaries – whether oceans, soils, forests, and atmosphere; people being torn from their lands and working long hours for a pittance or unemployed and desperate; or the future for our own and other species on this planet.

The consequences have become clear of attempts to mitigate and adapt to this crisis which do not question the basic premises of these models: when they neither accept the depth and extent of the linked technological, social, political and economic changes that will be required,[232] nor seek to enable the shifts in power relations necessary to allow solutions from the margins of the present system.[233] The appropriation of the climate challenge by neoli-beralism,[234] in particular the climate security agenda, is already having disturbing effects.[235] Its wider implications are terrifying, especially in the light of the abject failure of global political processes to make any meaningful progress towards genuine solutions.

In cheering – and often cheerful - contrast, numerous and uncounted experiments in commu-nity-based action towards constructive, enduring solutions offer significant glimmers of hope. Increasingly sophisticated in their philosophies, goals, practical approaches, decision-making processes and organisational structures, these emerging networks for collective action, infor-mation exchange and mutual support are growing into credible alternatives to a mainstream whose inability to address contemporary problems is ever more apparent.[236] At base the climate security agenda relies on people feeling fearful – preferably of some purported enemy, but if not then of the agenda drivers themselves. One crucial act of resistance is to refuse to enter that game, and to instead create our own. That is what these initiatives and social movements seek to do. In the process they are linking with and learning from enduring ways of improvising and sustaining effective and sustainable social and ecological relations among peoples on the margins of the global economy.

This chapter explores one of the largest and fastest growing (but still embryonic) of these movements, the Transition movement of grassroots efforts to cultivate community resilience to climate change and other symptoms of terminal dysfunction in the global economy. It seeks to place it in the broader context of commons-based ways of organising human expe-rience and society. In the process, it describes how Transition groups are actively visualising and realising economic alternatives within their own communities. Although their actions are as varied as the places themselves and people within them, they have in common that they seek to place the organisations and infrastructures that provide basic needs (food, energy and shelter) under the control of the people who depend on them. The realisation of this goal

---

232 Unruh, G., 2000. Understanding carbon lock-in. *Energy Policy* 28(12): 817-830.

233 Unruh, G., 2002. Escaping carbon lock-in. *Energy Policy* 30(4): 317-325.

234 Noble, D., 2007. *The Corporate Climate Coup*. *http://www.zcommunications.org/the-corporate-climate-coup-by-david-f-noble.* Accessed October 15 2012.

235 Buxton, N. & B. Hayes, 2015. *The Secure and the Dispossessed*. London: Pluto Press.

236 Hopkins, R., 2013. *The Power of Just Doing Stuff*. Cambridge: UIT/Green Books.

makes Transition and other similar efforts a powerful countermovement to responses that seek to entrench rather than transform the causes of crisis.

In this chapter, we examine Transition through the lens of commons: flexible and evolving institutional structures, often informal and/or customary in nature, through which the co-users of a shared resource experience, recognise and allocate rights and responsibilities. The neo-liberal appropriation of climate change has created new threats to these commons, including the enclosure of community forests under the REDD mechanism. We describe the efforts in North and South to reverse this enclosure and privatisation of resources by protecting existing commons regimes and creating new ones, as the basis for economic independence and communities' interdependence, and hence for community resilience cultural self-determination and mutual survival.

4.2.1. Transition: Building Resilience by Creating Economic Alternatives

The story the Transition movement often tells is that it started in 2006 as an informal cluster of volunteers in Totnes, a small market town in Devon, South West England, and that it has grown from this into an international movement that British green economist Tim Jackson has described as "the most vital social experiment of our time".[237] However, in reality, this work echoed and amplified movements that were already afoot in the Global North and South. For example, one of the first Transition initiatives in Scotland, in Portobello, Edinburgh, formed in 2005, when the label 'Transition' didn't exist. The group called itself Portobello Energy Descent and Land Reform Group (PEDAL) alluding to both the emphasis on building resilient communities that move away from oil addiction[238] and to the Scottish land reform movement's focus on reclaiming land (the fundamental source of our wealth and wellbeing) from the powers that be. Perhaps most crucially, Transition draws heavily on – and extends – theory and methods in permaculture, a design system for low-entropy human habitats that has spread worldwide since its origins in Australia in the 1970s and which mirrors many indigenous peoples ways of 'living with', rather than seeking to dominate, the environment on which they depend.[239]

People across the world have picked up on 'Transition' from their own historical contexts and for their own strategic reasons, partly because it offers a positive vision of the future and ways of acting in the here and now to bring that vision into being, in place of the dystopia being enacted in our name. When it works, it does so not as a blueprint imposed by those in the know, but as an adapting process of building resilience and connecting and learning from similar commons initiatives locally, regionally and internationally.

---

237 Hopkins, R., 2011. *The Transition Companion*. Totnes: Green Books.

238 This was picked up from Rob Hopkins' work with students at Kinsale in Ireland that led him to the idea of Transition *http://transitionculture.org/2005/11/24/kinsale-energy-descent-action-plan/* Accessed November 13th 2013.

239 Henfrey, T., & G. Penha-Lopes, 2015. *Permaculture and Climate Change Adaptation. Inspiring Ecological, Social, Economic and Cultural Responses for Resilience and Transformation*. East Meon: Permanent Publications. Pp. 27-29; 77-79.

The part of this commons movement we are identifying as Transition originated at Kinsale Further Education College in Ireland in 2005, when a lecture on peak oil inspired a group of permaculture students taught by Rob Hopkins to design a strategy for local independence from fossil fuels.[240] Impressed by the potential of this approach, in 2006 Hopkins relocated to Totnes, a South Devon market town with a longstanding reputation for countercultural action and co-founded Transition Town Totnes, extending the Kinsale approach and adding climate change as a second key concern.[241] Subsequently, in 2007, Transition Network came into being as a support and coordination body for a burgeoning number of Transition initiatives in other locations. As of September 2013, Transition Network reported the existence of 1130 local initiatives in 43 countries.[242]

Transition Network's website, like many of its other media, carries a 'cheerful disclaimer', which states:

> *"Just in case you were under the impression that Transition is a process defined by people*
> *who have all the answers, you need to be aware of a key fact.*
> *We truly don't know if this will work. Transition is a social experiment on a massive scale.*
> *What we are convinced of is this:*
> *› if we wait for the governments, it'll be too little, too late*
> *› if we act as individuals, it'll be too little*
> *› but if we act as communities, it might just be enough, just in time."*[243]

Transition is thus based on explicit scepticism about what top-down processes might achieve. It also acknowledges the limitations of acting solely as individuals, and instead focuses on the transformative potential of community action.

Transition's rapid growth in profile and popularity in large part resulted from its association with climate change at a time when the publication of the Stern Review put it at the heart of political agendas, both in the UK and internationally. At the time, peak oil received relatively little mainstream attention.[244] By explicitly linking the two, Transition approached climate change with a broader perspective that transcends the adaptation-mitigation distinction. Decarbonisation is viewed, not as an end in itself, but as a necessary condition for building

---

240 *http://transitionculture.org/2005/11/24/kinsale-energy-descent-action-plan/.* Accessed May 2nd 2014.

241 Hopkins, R., 2010. What Can Communities Do? In Heinberg & Lerch (eds.) *The Post Carbon Reader.* Santa Rosa: Post Carbon Institute.

242 *http://www.transitionnetwork.org/initiatives/.* Accessed November 13th 2013. A more recent blog post by Hopkins suggests this underestimates the numbers of non-UK initiatives, known nationally but not registered with Transition Network. *http://www.transitionnetwork.org/blogs/rob-hopkins/2014-04/impact-transition-numbers.* Accessed May 2nd 2014.

243 *http://www.transitionnetwork.org/support/what-transition-initiative.* Accessed October 15th 2012. Also see *http://transitionculture.org/2008/11/20/responding-to-greers-thoughts-on-premature-triumphalism/.* Accessed October 15th 2012.

244 An insider history of peak oil denial in the early 20th century can be found in Legget, J., 2013. *The Energy of Nations.* London: Earthscan.

Figure 4.2.1 – *Access to Land as a Prerequisite of Collective Action. Credit: Gesa Maschkowski.*

resilience, seen broadly as the capacity to negotiate change[245] and specifically as the ability to sustain provision of basic human needs in to the face of: shrinking supplies of cheap energy, the direct impacts of climate change, the consequences of necessary constraints on carbon emissions, and the economic instability to which all of these contribute.

The focus on resilience does not look at issues such as climate change mitigation, or energy security, in isolation, nor pretend they are well-defined problems amenable to simple solutions.[246] It highlights how, despite the immediate hardships and dangers they entail for many, these immediate issues are symptoms of deeper dysfunction in the economic system. This transforms the immediate need to address them into a longer-term opportunity to build resilience to future crises and in doing so address broader patterns of unsustainable resource use and inequalities of wealth and power.

245 Walker, B. and D. Salt, 2006. *Resilience Thinking: sustaining ecosystems and people in a changing world.* Washington DC: Island Press.

246 Prins, G. and S. Rayner, 2007. *The wrong trousers: radically rethinking climate policy.* Oxford: James Martin Institute for Science and Civilization. Prins, G. et al, 2010. *The Hartwell Paper: a new direction for climate policy after the crash of 2009. http://eprints.lse.ac.uk/27939/*

BOX 1: TRANSITION AND RESILIENCE

There is a great deal of inconsistency and confusion about the usage and meaning of the term 'resilience' in both academic and non-academic settings. In particular, the meaning and implications of the term in Transition are very different from those in prominent policy discourses associated with securitisation.

Transition draws on insights originating in ecology[247] that treat resilience (and hence sustainability) as dynamic conditions depending as much on flexibility as stability: on adapt to rather than resisting change.[248] Nick Wilding observes that community resilience, in this framing, is an ongoing condition: while it may be most apparent at times of crisis, it is built and maintained by the ongoing development of relationships, social and environmental knowledge, and capacities for collective action.[249]

This contrasts with uses of the term in risk management and disaster response literatures, which emphasise the need to maintain or to return to some pre-existing, presumably desirable state following disturbance. Their framing engages only superficially, if at all, with established technical definitions of social-ecological resilience.[250] As some critical commentators have observed, this usage has passed uncritically into popular and policy discourses that treat personal and community resilience as panaceas for the corrosive effects of predatory capitalism.[251]

Ignoring political, ecological and dynamic dimensions of the definition and deployment of resilience invites confusion between its meanings within Transition and its use as a tool for normalising the effects of politically conservative agendas.[252] For Transition, it is a transformative concept, founded on the understanding that our economic system is already operating beyond finite ecological boundaries and that resilience must embrace transformation: approached through a combination of practical and inner work that subverts the addiction to commodities as a substitute for meaningful relationships and replaces it with creative systems that renew social, cultural and material relations.[253]

247 Haxeltine, A., & G. Seyfang, 2009. *Transitions for the People: theory and practice of 'Transition' and 'Resilience' in the UK's Transition movement.* Tyndall Centre for Climate Change Research Working Paper 134. Hopkins, R., 2010. *Localisation and Resilience at the Local Level: the Case of Transition Town Totnes (Devon, UK).* Ph.D. thesis, Plymouth University.

248 Leach, M., I. Scoones & A. Stirling, 2010. *Dynamic Sustainabilities.* London: Earthscan.

249 Wilding, N., 2011. *Exploring Community Resilience in times of Rapid Change.* Dunfermline: Carnegie UK Trust. http://www.carnegieuktrust.org.uk

250 Alexander, D. E., 2013. Resilience and disaster risk reduction: an etymological journey. *Natural Hazards and Earth Systems Science* 13: 2707–2716.

251 Neocleous, M., 2013. Resisting Resilience. *Radical Philosophy 178.* http://www.radicalphilosophy.com/commentary/resisting-resilience. Accessed November 27th 2013.

Following the economic crisis of 2008, Transition adopted the imperative of a post-growth economy as its 'third driver'. New audiences have become receptive to arguments that highlight the non-viability of the global economy, and links with organisations like the New Economics Foundation - an important influence on Transition since its earliest days - have been consolidated. Practically, it has led to an emphasis on social enterprise as a tool for provision of basic needs and livelihood creation not reliant on the fossil fuel economy[254]. This entrepreneurial approach to sustaining Transition projects could be viewed as a partial sell-out - working within, and therefore perpetuating, the existing system rather than challenging its basic values – or as a deeply subversive form of head-on confrontation. Either way, the embedding – or making explicit – of an economic critique within the central tenets of Transition is perhaps a sign of new maturity.

For many in Transition, the fundamental weaknesses of our global economy have also been challenging to accept. In a workshop at the 2010 Transition Network conference, Stoneleigh, an influential blogger on economic crisis, presented a powerful vision of imminent collapse in the global economic system and the consequences of this. Many participants found this message profoundly disturbing – to the extent that organisers changed the planned programme to include a public airing of views and concerns over the issue. Stoneleigh's specific predictions have not borne out: through mechanisms like quantitative easing, the global financial system has shown remarkably obduracy, at least in the short term (and notwithstanding the prospect that such 'band-aid' measures will worsen the long-term consequences). But the effect of this information on an audience of committed, seasoned, well-informed climate activists shows the depth of our psychological and cultural attachments to the conditions associated with constant economic growth.

Janis Dickinson has written about climate change denial as an immortality project: the truth about climate change, she argues, is so shocking, so unthinkable, that to accept it implies a shattering of identity too difficult for many to bear.[255] Transition's emphasis on 'Heart and Soul': the 'Inner Transition' demanded by the needs to accept the loss of the familiar, to assume personal responsibility to take action, and for ongoing emotional support when experiencing these, is in great part based on recognition of this. Perhaps even greater trauma lies in acknowledging the need for radical transformation in our economic system since this

252 MacKinnon, D., & K. D. Derickson, 2012. From resilience to resourcefulness: A critique of resilience policy and activism. *Progress in Human Geography* 37(2): 253-270. Cote, M., & A. J. Nightingale, 2012. Resilience thinking meets social theory: Situating social change in socio-ecological systems (SES) research. *Progress in Human Geography* 36(4): 475-489.

253 Bailey, I., R. Hopkins & G. Wilson, 2010. Some things old, some things new: The spatial representations and politics of change of the peak oil relocalisation movement. *Geoforum* 41(4): 595-605. Brangwyn, B., pers. comm., Nov 28th 2013. Henfrey & Giangrande, this volume.

254 http://www.reconomyproject.org/

255 Dickinson, J. L. 2009. The people paradox: self-esteem striving, immortality ideologies, and human response to climate change. *Ecology and Society* 14(1): 34. [online] URL: http://www.ecologyandsociety.org/vol14/iss1/art34/ Accessed October 15 2012. Also see Norgaard, K.M., 2011. *Living in Denial: Climate Change, Emotions, and Everyday Life.* Cambridge, MA: MIT Press.

requires not just acceptance of the possibility of drastic (but for now and for those in the affluent global north, still perceived as somehow distant and delayed) changes in the future, but of change in our material conditions of existence right now.

Translated to the stage of international politics, this immortality project manifests itself in programmes that, despite the evidence to the contrary, present economic growth as the solution to climate change rather than as an underlying cause of it. In part this is due to perceived vested interests, whether of us as individuals, of the majority in Western society, or of the wealthiest one percent in it. For while increased inequality is to everyone's disadvantage,[256] only in a growing economy can the rich and powerful increase their wealth while simultaneously increasing others' access to materials goods, or increasing spending on social control, in sufficient measure to diffuse or control dissent.[257] Such denial also reflects macro-economic constraints. Economies in which money is created largely through debt are only stable under conditions of growth.[258] More fundamentally, it reflects crises of belief and imagination: a world without economic growth has been made quite literally unthinkable not just to economic elites, but to the majority of ordinary people whose livelihoods depend on participation in that economic system.[259]

> *Translated to the stage of international politics, this immortality project manifests itself in programmes that, despite the evidence to the contrary, present economic growth as the solution to climate change rather than as an underlying cause of it.*

Carbon lock-in, as Gregory Unruh termed our systemic addiction to fossil fuels, is more than the set of linked technical, institutional and political interdependencies he described. It is a cultural phenomenon - an incredibly deep-rooted one - resting upon a political economy that has achieved a status equivalent to religion, in terms of the dogmatic faith of its protagonists, and in its material necessity for followers who are rendered uncritical through 'education', repetition, and (effectively) indoctrination.[260] These linked conceptual and structural depend-

256 Wilkinson, R. & K. Pickett, 2009. *The Spirit Level*. London: Penguin.

257 Douthwaite, R., 1999. *The Growth Illusion*. Second edition, revised. Totnes: Green Books.

258 Jackson, T., 2009. *Prosperity Without Growth*. London: Earthscan.

259 Eisenstein, C., 2011. *Sacred Economics*. Berkeley: Evolver Editions.

260 Jackson, T., 2006. Consuming Paradise? Towards a socio-cultural psychology of Sustainable Consumption. In Jackson, T. (ed.), *Earthscan Reader in Sustainable Consumption*. London: Earthscan. Filk, R., 2009. Consuming ourselves to death: the anthropology of consumer culture and climate change. *In* Crate, S.A. & M. Nuttall, 2009, *Anthropology and Climate Change: from Encounters to Actions*. Walnut Creek: Left Coast Press.

encies on growth are an example of what Anne Wilson Schaef has characterised as process addictions, manifest at the level of society as a whole.[261] In her analysis, everyone within this society to some degree displays and suffers from personal symptoms of this addiction: including the recovering addicts who recognise and oppose the current state of affairs. Transition has adopted this language of addiction and recovery as a central concept, and the need for inner change - both individual/psychological and collective/cultural - as a core principle.

Inner transition is perhaps what more than anything sets Transition aside from other forms of environmental and social activism with a more single-minded emphasis on the material and practical aspects of change. Transition makes the use of the imagination central to its practical method, particularly through the use of backcasting.[262] This invites participants to imagine a more positive future for their community, free of dependency upon fossil fuels, and work back from there to identify the immediate steps necessary to realise it. This is a powerful antidote to capitalism's control over the imagination. It frees the imagination to set goals and measures that inevitably prove incompatible with capitalism's basic premises: community-scale projects of a variety of forms, institutionalised in inclusive and democratic structures of ownership and decision-making.

In practical terms, the projects, organisational forms and internal cultures of Transition initiatives represent an exercise in creating new commons. While the creation of commons is innovative in this context, it is also the rediscovery of a set of principles that have been in active use for millennia, and in direct opposition to the expansion of capitalism since its very origins. It thus links Transition and other movements for building economic and social alternatives with the struggles of peoples marginal to the global economy who are already organising their economic lives around common property regimes. This emergent global commons movement represents a collective grassroots response to climate change incompatible, both ideologically and materially, with the climate securitisation agenda.

4.2.2. Climate and the Defence of Commons

One way of understanding commons is as the relationships that constitute place, and the care we need to take to ensure that all (human and non- human) aspects of this place flourish. People in the Global North may understand this approach best as the attitude we bring to being at home. One Ogiek man from Mt Elgon in Kenya managed to communicate this depth of relationship to place by describing how when their community was forced off their common lands and forced to live elsewhere it felt to him like being forced to leave his wife and children and being given a different family.

A more technical understanding of commons is that they are land or other resources under the effective ownership of groups of co-users, who manage them collectively by creating and implementing agreed rules of access and appropriate usage, monitoring actual behaviour

261 Wilson Schaef, A., 1987. *When Society Becomes an Addict*. San Fransisco: Harper Row.
262 *http://www.thenaturalstep.org/backcasting*, Accessed October 15th 2012.

for conformity with these rules, enforcing appropriate sanctions when they are broken, and changing them in the light of experience.[263]

Globally, the scale of common land is vast despite successive waves of privatisation and dispossession, states often seeing commons as public lands for the government to dispose of as it wishes. For example, in sub-Saharan Africa communities hold around 1.6 billion hectares under customary laws, around 75 percent of the total land area.[264] These global commons embrace a vast range of land management regimes, based on complex collective decision-making processes subtly adapted to specific local ecological, social and cultural conditions.[265] This matching of institutional and ecological diversity, known as biocultural diversity, is a vital aspect of human adaptation to different habitats.[266] Common property regimes are in every documented case a key element of sustainability and resilience in resource use.[267]

Many indigenous populations have historical experience of climatic change[268] and the negotiation of extreme weather events.[269] The decision-making procedures associated with their common property institutions both enable and reflect collective learning from these events. The flexibility of these regimes in the face of changing or unpredictable conditions is an essential aspect of the people's adaptability, and hence the resilience of the social-ecological systems of which they are a part.[270] Each is a set of rules and procedures fine-tuned to the details of its social, cultural and environmental context that incorporates mechanisms for self-evaluation and adjustment in responses to changes in this local context. These localised commons and the global biocultural diversity they support represent a collective cultural commons that will be humanity's main source of ideas and knowledge serving to help us negotiate climate change and other major environmental disturbances.

263 Ostrom, E., 1990. *Governing the Commons: The Evolution of Institutions for Collective Action*. Cambridge University Press. Bromley, D.W. (ed.), 1992. *Making the Commons Work*. San Fransisco: Institute for Contemporary Studies.

264 Alden Wily, L., 2011. 'The Law is to Blame' Taking a Hard Look at the Vulnerable Status of Customary Land Rights in Africa. *Development and Change* 42(3).

265 Ostrom, E., 2005. *Understanding Institutional Diversity*. Princeton University Press.

266 Maffi, L. (ed.) 2001. *On Biocultural Diversity*. Washington and London: Smithsonian Institution Press.

267 Berkes, F. (ed.), 1990. *Common Property Resources*. Berkes, F. & C. Folke (eds.), 1998. *Linking Social and Ecological Systems. Management practices and social mechanisms for building resilience*. Cambridge: Cambridge University Press.

268 Grove, R.H., 1997. *Ecology, Climate and Empire*. Cambridge: White Horse Press.

269 Firth, R., 1959. Critical Pressures on Food Supply and their Economic Effects. *In Social Change in Tikopia*. London: Allen and Unwin. Waddell, E., 1975. How the Enga Cope with Frost: Climatic Perturbations in the Central Highlands of New Guinea. *Human Ecology* 3(4): 249-273. Both reprinted in Dove, M.R., and C. Carpenter (eds.), 2008. *Environmental Anthropology: a Historical Reader*. Oxford: Blackwell.

270 Berkes, F. 2008. *Sacred Ecology: Traditional Ecological Knowledge and Resource Management*. Second edition, revised. London: Routledge. Salick, J., & A. Byg, 2007. *Indigenous Peoples and Climate Change*. Oxford: Tyndall Centre.

Mainstream responses to climate change that perpetuate capitalism's onslaught against both specific commons and the global cultural commons emergent upon them, systematically undermine our capacity as a species to respond in other, more constructive and effective ways. An instructive example is Aboriginal Australian fire burning regimes. These create highly ecologically variegated

> *Localised commons and the global biocultural diversity they support represent a collective cultural commons that will be humanity's main source of ideas and knowledge serving to help us negotiate climate change and other major environmental disturbances.*

landscapes that support higher biodiversity and reduce the likelihood of large-scale fires. When they were banned by the authorities, biodiversity plummeted as landscape matured to a uniform condition, which also allowed fires to spread unabated. In 2009, corporations began paying Aboriginal rangers millions of dollars to re-establish their fire regime practices, through the West Arnhem Land Fire Abatement Project. In its first year the areas susceptible to burning reduced from a previous average of 37 percent to only 16 percent of forest. Over the first four years this generated what the project described as a saving of 488,000 tons of carbon dioxide.[271] Of course people in commons regimes would simply never abstract out singular benefits like this; for it is this very pattern of treating elements in isolation (as markets are prone to do) that has led to the social and ecological disasters generated by the enclosure and destruction of commons-based regimes.

Current scientific understanding of the role of commons is markedly at odds with historical and even contemporary attitudes prevalent in capitalist societies. The latter often refer to Hardin's notorious 'tragedy of the commons',[272] although Hardin himself is reported later to have acknowledged that he had confused commons with open-access regimes, or unregulated commons.[273] In reality, whether in relation to a stretch of river or coast for fishing, or to an area of forest or grazing land, a prerequisite for the development of a common property regime is very clear **demarcation** of who can and cannot use the resource, and on what terms. On this basis, intricate and flexible systems of rights and responsibilities evolve, including mechanisms for **incorporating** new co-users of the common pool resource who are willing to abide by the **reciprocal rules** required to maintain it.[274]

271 http://savanna.cdu.edu.au/information/arnhem_fire_project.html. Accessed May 3rd 2014.

272 Hardin, G., 1968. The Tragedy of the Commons. *Science* 280: 682-683.

273 Kirkby, J., P. O'Keefe & L. Timberlake (eds.), 1995 'The Commons: where the community has authority' in *The Earthscan eader in sustainable development*. London: Earthscan.

274 Ostrom, E., 1990. *Governing the Commons: The Evolution of Institutions for Collective Action*. Cambridge University Press. Kenrick, J., 2006. 'Equalising Processes, Processes of Discrimination and the Forest People of Central Africa', in Widlock, T. & W Tadesse (eds.) *Property and Equality: Vol. 2 Encapsulation, Commercialization, Discrimination*. Oxford: Berghahn.

The self-organising nature of commons, and their consequent need for non-interference by external authorities, has often led states and other powerful actors to undermine them: sometime purposefully, sometimes through ignorance.[275] Removal of commons regimes serves several purposes. Denial of the existence of other ways of organising bolsters the ideological power of states and markets. Appropriation of land for other (profit-generating) purposes and the people on it as sources of cheap labour both boosts industrial systems and makes land and people more tractable to central regulation and control.

> *The self-organising nature of commons, and their consequent need for non-interference by external authorities, has often led states and other powerful actors to undermine them*

This transfer of people, goods and services from the common pool into the monetary realm, initially via enclosures and land clearances in seventeenth century Britain, was central to capitalism's early establishment and subsequent, ongoing, worldwide expansion.[276] A Yarralin Aboriginal man, Hobbles, described how following the massacre of over 95 percent of Yarralin people in the Victoria River Delta between 1883 and 1939, the survivors were forced into slave labour at the cattle stations established on their land.[277] It continues apace as corporations supported by states take land and other resources from those whose commons regimes still sustain livelihoods based on farming, hunting, fishing and/or foraging.[278]

However, commons systems rarely completely disappear, and can often be revived or enhanced. Box 2 describes how Ogiek people in Kenya are resisting eviction from their ancestral lands by documenting and formalising customary laws and practices, and making these the basis of cooperation with state authorities who share their concern for protection of the resource base.

275 Scott, J.C., 1998. *Seeing Like a State: How Certain Schemes to Omprove the Human Condition have Failed.* New Haven and London: Yale University Press. Graeber, D., 2004. *Fragments of an Anarchist Anthropology.* Chicago: Prickly Paradigm Press.

276 Fairlie, S., 2009. A short history of enclosure in Britain. *The Land* 7: 16-31. Tate, W.E., 1967. *The English Village Community and the Enclosure Movements.* London: Victor Gollancz.

277 Rose, D., 2000. *Dingo Makes Us Human.* Cambridge University Press.

278 Geary, K., 2012. *Our Land: Our Lives.* Oxfam Briefing Note. Oxford: Oxfam International.

BOX 2: COMMONS REGIMES AND COMMUNITY EMPOWERMENT IN KENYA

The Ogiek of Mount Elgon in Kenya have been evicted many times from their ancestral lands at Chepkitale. They have recently put into writing the bye-laws which govern their commons, among other reasons to explain to conservationists who have sought to expropriate their land that the Ogiek themselves are best placed to ensure the land's well-being. They explain that:

> *"We have never conserved. It is the way we live that conserves. These customary*
> *bye-laws we have had forever, but we have not written them down until now."*

The bye-laws were finalised in an intense community process of mapping and dialogue. In a sense, they simply document how the Ogiek have organised themselves and managed their lands since time immemorial, but as one community member pointed out, "When you write things to say this is what we should do then you get community members who disagree and you have to decide what to do."

For example, the bye-laws banned charcoal burning, prompting passionate debate: especially by those who hadn't attended meetings. Several people asked, "Why should we protect the forest when others are expanding their fields into it and burning it to sell as charcoal? Why shouldn't we also benefit from forest destruction?" In subsequent dialogue the overwhelming majority agreed the forests are vital as cattle pastures and for beekeeping. Their conservation depended on majority agreement not to destroy a common resource for short-term advantage and requiring dissenters to abide by the consensus or face sanctions.

The Ogiek next sought to inform various authorities of these bye-laws and to seek their support in implementing them. The District Commissioner applauded the community for being stronger on conservation matters than any other authority. When Ogiek community scouts began to arrest illegal charcoal burners, this prompted the Kenya Forest Service to do the same.

Unfortunately, this insightful community is simultaneously involved in an ongoing legal struggle to reclaim lands gazetted by the Government in 2000, making the Ogiek living there 'illegal trespassers'.[279] The Ogiek are hoping the Government will settle out of court: acknowledge Ogiek ownership of their commons and work hand in hand with the community to demonstrate that human rights based conservation is a new way of explaining an age old idea that if you look after the land it will look after you.

279 *http://whakatane-mechanism.org/kenya.* Accessed May 1st 2014.

With dark irony, excuses for these land thefts are often framed in terms of poverty alleviation. Proponents draw attention to rises in monetary income, often from close to zero, for the dispossessed - who have had no other recourse for survival than to seek poorly paid wage labour. However, purely fiscal analysis obscures the destruction of their real wealth, which was the ability to provide for their families sustainably while having the option to engage in monetary transactions to cover essential costs such as school fees. Speaking in September 2012 to Vai communities whose farmlands, rice swamps, fishing creeks and sacred forests had been bulldozed by a Malaysian palm oil company, the President of the Liberian Senate said, "Our actual intention is to change your lifestyle from farmers to workers so you no longer grow cassava or rice but work for money to buy rice and cassava that has been grown by someone else... We say to you: no longer will you have land to grow rice, cassava and peppers."

4.2.3. Community Ownership or REDD? Contrasting Commons and Commodity Solutions to Deforestation

Another clear example of the contrasts between attempts to solve global problems that seek respectively to protect/extend commons and to enclose them is evident in approaches to protecting the forests of the Global South. A commons approach would protect forests through ensuring protection of the rights of local communities, and hence through protecting their management practices and their stake in the forest as not just a resource but also as home. The politically dominant approach is based on the assumption that the best way to reduce rates of deforestation is through putting forests under the control of large conservation bodies, corporations and governments, for example through REDD projects that seek to make the carbon in those forests a tradable commodity.

The quantitative evidence points to the efficacy of a commons-based approach. Comparison of 40 protected areas and 33 community-managed forests in 16 countries across Latin America, Africa and Asia showed average annual deforestation rates in protected areas were six times higher than in forest managed by local communities.[280] Research for the World Bank's Independent Evaluation Group found that: "In Latin America, where indigenous areas can be identified, they are found to have extremely large impacts on reducing deforestation".[281] For example, Brazil has a large network of indigenous territories representing 20 percent of the legal Amazon. Only 1.3 percent of total deforestation in the Amazon occurs inside these territories, which are 98.4 percent preserved. In contrast, government ownership of forests is associated with unsustainable forest use. This is because when local users perceive insecurity in their rights (because the central government owns the forest land), they seek to maximise short-term livelihood benefits through fear they will lose these benefits to others.

Part of the reason why REDD was thought to be a cheap option in the fight against climate chaos,

280   _http://www.cifor.org/online-library/browse/view-publication/publication/3461.html._ Accessed October 15th 2012.

281   Nelson A. & K.M. Chomitz, 2011. Effectiveness of Strict vs. Multiple Use Protected Areas in Reducing Tropical Forest Fires: A Global Analysis Using Matching Methods. _PLoS ONE_ 6(8): e22722. doi:10.1371/journal.pone.0022722.

was because it was seen as – in effect – being about controlling poor people's behaviour. If the drivers of deforestation are recognised as being the large players then there is a very different economics. Swedish research has shown that payments for abstaining from converting forests to – for example – oil palm plantations simply cannot reach a level that would make this alternative more profitable than the plantation. Clearly, the cheapest, most efficient and most reliable way to protect forests is to support the recognition of customary land tenure and the protection of sustainable land use in the forests of the Global South.

Depending on how it is implemented, REDD+ therefore presents a real danger of promoting a 'fortress conservation' approach that further destroys commons by excluding and marginalising forest peoples, at the same time as it can potentially provide opportunities for the recognition of rights, and securing of community forests, through international scrutiny and national tenure reform.

Marginal peoples have not been passive in the face of threats of this type. Wapichan people in southern Guyana, awarded title to only part of the land area claimed in a petition to the national Amerindian Lands Commission in 1969,[282] recently released a comprehensive survey of their use of the area, reasserting their original claim on the grounds of forest stewardship and UN recognition of people's right to self-determination.[283]

4.2.4. Transition, Tradition, and Rediscovery of the Commons

As touched upon above, the term 'the commons' can refer to a far broader range of resources and associated activities than just use of land and natural resources. Kenrick has previously distinguished **community commons** including natural resources and shared community institutions (such as those for resource allocation and dispute settlement, child care and care for the elderly, health care and community provided education) from **cultural commons** such as literature, music, arts, design, film, video, television, radio, information, open source software and collectively created and maintained internet resources such as Wikipedia.[284] Far more so than community commons, cultural commons co-exist with capitalism in a range of complicated interrelationships, perhaps because the industries involved are more sharply aware that they are sources of flexibility and creativity that the corporate world itself can not match, but that are essential for its survival in a fast-changing world.[285] One of the fatal contradictions at the heart of capitalism is that it is reliant on both community commons and cultural commons. Many of its costs are externalised upon environments that consist entirely of communities of relationships, others onto people's home lives, non-monetary exchanges,

---

282 Henfrey, T., 1999. Land Conflicts and Cultural Change in Southern Guyana. Pp. 328-333 in Grenand, P. & F. Grenand (eds.), 1999. *Les Peuples des Forêts Tropicales Aujord'hui. Volume IV: Volume Regional Caraibes.* Brussels: EC DG8.

283 http://www.forestpeoples.org/topics/customary-sustainable-use/publication/2012/wapichan-people-guyana-make-community-based-agreem. Access October 15th 2012.

284 Kenrick, J., 2012. The Climate and the Commons. In Davey, B. (ed.), *Sharing for Survival*. Cloughjordan: FEASTA. http://www.sharingforsurvival.org/index.php/chapter-2-the-climate-and-the-commons/.

285 Leadbeater, C., 2008. *We-Think*. London: Profile Books.

and the emotional support that enables people to continue operating in hierarchical work places.

The importance to Transition of both community and cultural commons is deep-seated. It builds upon long-standing recognition of the importance of cultural creativity in radical social movements,[286] and radical environmentalism in particular.[287] It also draws upon Transition's roots in permaculture, which emphasises the importance of designing productive habitats and associated management regimes in ways that reflect the unique details of the local ecology.[288] This place-specificity naturally extends to human dimensions such as social institutions, cultural practices, and personal histories,[289] and how these are reflected in the built and residential

Permaculture's growth as a movement can thus be considered a global grassroots experiment in the deliberate creation of biocultural diversity.

environment.[290] Permaculture's growth as a movement can thus be considered a global grassroots experiment in the deliberate creation of biocultural diversity.

As a 'feral ecology',[291] permaculture democratises the interpretation and application of ecology far beyond the conventional limits of science. It does this largely through the creation of cultural commons that reflect the movement's collective understanding about how to organise society along ecological principles in ways that reflect its basic ethical commitments to environmental and social justice[292]. These cultural commons include literature (much of it distributed freely as electronic media, regardless of its copyright status), formal and informal organisational structures, pedagogy, customs, social networks, and practitioners' knowledge. To the extent that it makes sense to differentiate Transition from permaculture, it is perhaps in

286 McKay, G., 1996. *Senseless Acts of Beauty: Cultures of Resistance Since the Sixties*. London: Verso.

287 Butler, B., 1996. The tree, the tower, and the shaman: the material culture of resistance of the No M11 Link Roads Protest of Wanstead and Leytonstone. *Journal of Material Culture* 1(3): 337-363. Guattari, F., 2000 (1989). *The Three Ecologies*. Translated by I. Pindar and P. Sutton. London: Athlone.

288 Mollison, B., 1988. *Permaculture: a Designer's Manual*. Tagari Publications. Whitefield, P., 2009. *The Living Landscape*. East Meon: Permanent Publications.

289 Macnamara, L., 2012. *People and Permaculture*. East Meon: Permanent Publications.

290 Alexander, C., 1978. *A Pattern Language*. Berkeley: Center for Environmental Structure. Pickerill, J., 2013. Permaculture in practice: low impact development in Britain. In Lockyer, J., & J. Veteo (eds.), *Localizing Environmental Anthropology: Bioregionalism, Permaculture, and Ecovillage Design for a Sustainable Future*. New York: Berghahn Books.

291 Morris, F.A., 2012. When Ecology Goes Feral. *Wageningen Journal of Life Sciences* 59: 7-9.

292 The three core ethics of permaculture are most commonly referred to as 'Earth Care', 'People Care', and Fair Shares. E.g. see Macnamara, L., 2012. *People and Permaculture*. East Meon: Permanent Publications. Pp. 4-7.

how Transition stresses how peak oil and climate change give this project increased urgency and impetus. The main practical consequence – aside from perhaps more emphasis on urban as well as rural action[293] - is more systematic attention to the creation of new cultural and community com mons as a core method and goal.

Transition initiatives are as distinctive and varied as the communities in which they are based, but their development tends to exhibit a number of common 'patterns' or 'ingredients'.[294] A typical initial focus is on creating common pools: of shared understanding and knowledge through awareness-raising; of basic infrastructure for communication through harvesting email addresses and setting up mailing lists and websites (themselves usually based on open source platforms, among the most important recent cultural commons), identifying suitable meeting and event spaces, and agreeing processes for organising meetings, reaching decisions, allocating responsibilities, and welcoming new members to the group. A group may later formalise these (although many choose not to), when it adopts a written constitution or legally registers as an organisation. Other early activity – which may have preceded organisational development, or follow or run alongside it - often focuses on creating or improving physical community commons through visible manifestations such as gardens, orchards, renewable energy projects, community bakeries and other enterprises, which add value to existing community spaces or bring new ones into being. Further activity – at present representing Transition's cutting edge - extends and deepens this physical common pool, for example by creating complementary currencies; establishing renewable energy co-ops with membership open to everyone in the community; initiating new community-based social enterprises; and securing access to land for alternative forms of food production rooted in permaculture and agroecology, housing, business premises and other forms of new low carbon community infrastructure.

Innovation may take place anywhere in the network. As might be expected, Transition Town Totnes has pioneered much activity, and along with Hereford and Brixton coordinates national work on the economic implications of localisation. Portobello Energy Descent and Land Reform Group, in addition to its work on land reform, has developed a community market, an orchard and community renewables. Community arts are central in the activities of Transition Town Tooting in South London, a source of inspiration to Transition initiatives elsewhere. Transition Norwich has pioneered work on low carbon lifestyles – connecting people who are committed to making the changes immediately in their own lives, for example by committing to not buying anything new or not owning a private car. Members of Transition Liverpool and Transition Durham with professional backgrounds in academic research set up mechanisms to improve connections with academic researchers.[295] Involvement of Transition

293 Lockyer, J., 2010. Intentional community carbon reduction and climate change action: from ecovillages to transition towns. In M. Peters, S. Fudge, T. Jackson (eds.) *Low Carbon Communities: Imaginative Approaches to Combating Climate Change Locally.* Cheltenham: Edward Elgar.

294 Hopkins, R., 2011. *The Transition Companion.* Totnes: Green Books.

295 *http://www.transitionresearchnetwork.org*

Figure 4.2.2 – *Celebration as an Integral Part of Community Building. Credit: Gesa Maschkowski.*

groups in community energy, with Bath, Brixton and Lewes particularly prominent, has often drawn on resources developed outside the movement itself.

As Transition has grown into an international movement, local responses to the very different conditions experienced outside the UK have taken very different shapes (Box 3). This diversity increases the range of responses available to all Transition groups to the likely consequences of climate change and further economic crisis. Transition Network, along with various national and regional hubs, coordinates the exchange of news and information about these activities via its website and other communications mechanisms, and organises collaborative events and projects such as Reconomy, which involves several initiatives, projects and organisations in the UK and internationally and examines relationships between transition and business.

BOX 3: INTERNATIONAL DIVERSITY OF TRANSITION

Transition has become an international movement, and very different approaches have emerged to reflect local and regional circumstances. In Portugal, for example, many people and organisations have experienced the effects of the financial crisis as a situation of 'peak money'.[296] The scarcity of money is regarded as both a situation to which it is necessary to adapt and an opportunity for innovation.

Filipa Pimentel of Transition Network has described how Portuguese Transition groups enact the idea of the 'gift economy',[297] organising events on zero financial budgets and without requiring either cash donations from participants or external funding. This approach obliges groups to rely on their existing skills, knowledge and resources, highlights and helps strengthen these, and additionally ensures local provisioning of both material goods and non-material assets. Practical activities include the regeneration of neglected sites owned by public bodies such as universities and local authorities as community spaces for leisure and food production. The DIY, participatory approach is creating new knowledge commons, for example at the Ajudada event in June 2013, which assembled over 450 people from all walks of life to discuss what a people-centred economy would look like and how to make it happen.[298]

In Brazil, Transition first took root in favelas, slum areas on the margins of major cities such as Brasilandia in Brasilia. In these cash-poor communities, issues such as food security and diet-related nutritional deficiencies, violence, and access to basic health and educational services are current and major concerns. Solutions include mapping open spaces in the city and turning them into community gardens where fruit and vegetables are grown for consumption within the neighbourhood, barter markets, a community bakery, and 'upcycling' businesses making bags out of old advertising banners.[299]

One might wonder what all these initiatives have to do with anticipating and helping plan for times of climate crisis? The key is that they are focused on building relationships of place in the present. The climate is rarely the central focus: such a focus on a devastating problem tends to paralyse. Rather the focus is on how the community can restore the commons. Put another way: how can the community kick the habit of consumption and competition that is promoted as the only game in town?

296 *http://transitionculture.org/2012/04/27/a-report-on-peak-money-and-economic-resilience-a-transition-network-one-day-conversation/.* Accessed November 27th 2013.

297 *http://transitionculture.org/2012/05/08/filipa-pimentel-on-transition-in-portugal-we-try-to-reduce-money-exchange-in-everything-we-do/.* Accessed November 27th 2013.

298 *http://www.ajudada.org.* Accessed November 27th 2013.

299 *http://transitionculture.org/2013/02/07/what-transition-looks-like-in-brazil/.* Accessed November 27th 2013.
Hopkins, R., 2013, *The Power of Just Doing Stuff*, pp. 113-4.

Experiences of Transition initiatives show the links between the creation of new commons and long-term projects of building resilience. This is sometimes true even where local groups have apparently lost momentum or subsided, and particularly common in places where economic contraction has significantly affected local livelihoods. One Transition initiative in mid-Wales set up a market stall for domestic vegetable growers with excess produce, which acts as a common pool trading point. The quantity of trade grew greatly after 2009, when increasing numbers of people began selling produce to compensate

The creation, extension and cultivation of community and cultural commons also provides a link between Transition's focus on local action and activity at broader scales.

for unemployment or reduced incomes, and buyers experienced more reliable supplies and stable prices. At the same time, a garden share scheme transformed private gardens into common growing spaces, allowing greater numbers of people to produce for both home use and sale.

Elsewhere, car share schemes – transferring vehicles from private to semi-public status – have proven vital in maintaining acceptable levels of mobility in rural areas with limited public transport provision. Energy efficiency measures and switching from oil heating to locally sourced woodfuel have insulated people from the effects of rising oil prices, based on a cultural commons of knowledge about technologies and their uses.[300] All of these are examples where people have created new commons in order to build resilience to peak oil; all have later proven sources of resilience against economic instability.

4.2.5. Beyond Localisation: Transition and Global Climate Justice

The creation, extension and cultivation of community and cultural commons also provides a link between Transition's focus on local action and activity at broader scales. Although much Transition-related discourse equates resilience with localisation,[301] true resilience is a product of interactions among functioning structures at multiple scales:[302] at broad scales emergent

---

300 Examples in this paragraph were reported by participants in an open space session entitled, 'Are small and slow solutions resilient in extraordinary times' at the Transition Network conference, Battersea Arts Centre, London, on 15th September 2012 – thanks to them for sharing and giving permission to report on them here.

301 Haxeltine, A., & G. Seyfang, 2009. *Transitions for the People: theory and practice of 'Transition' and 'Resilience' in the UK's Transition movement.* Tyndall Centre for Climate Change Research Working Paper 134.

302 Gunderson, L. & C.S. Holling (eds.), 2002. Sustainability and panarchies. Pp. 63-102 in Gunderson and Holling (eds.) *Panarchy.* Washington DC: Island Press.

upon diverse local situations,[303] and at local levels reliant on larger scales providing suitable contexts for adaptation to change.[304] If the prosperity of one locality is predicated on undermining resilience elsewhere, it is not in fact resilient to changes in the conditions of political economy, and indeed ecology, that allow such exploitation.

Transition's hopeful view of possible futures thus depends on the belief that we can only achieve them if we move towards them together. It is increasingly obvious that we also need to know what we are up against: that in some sense we are trying to create islands of cooperation in a viciously competitive system. Hope and political awareness are thus in synergy: the best way to motivate people to work together to create new commons is to focus on the positive and immediate benefits of this collective action: both within communities of place and through networks of solidarity and cooperation at broader scales.

The global economic system's current fragility due to its dependence on fiscal growth and externalised environmental and social costs is becoming ever more apparent as global limits are reached. For this reason, the current emphasis within Transition on localisation makes more sense viewed as a necessary corrective to excess globalisation than an end in itself. Transitioned communities could not long survive against broader backgrounds of climate chaos and conflicts arising from uneven distribution of resources and the human capacities to make use of them. Their resilience depends not only on the properties of localised production systems, but on emergent capacities to buffer variation in these through material, intellectual and cultural interchange and other forms of mutual aid. A community of any size is only as resilient as its nearest neighbour, which is one reason Transition was conceived as a replicable model. The Transition vision is not, and never has been, one of gated eco-communities isolated from the wider world, but is one of maximum local self-reliance as a basis for solidarity and cooperation, and identifying the appropriate scales for productive activities not feasible at local levels.[305] As a movement, it is neither discrete nor well-defined, but overlaps, intersects and links with numerous others. It contributes to creating and strengthening cultural and community commons far greater in significance and extent than its own efforts could achieve in isolation.

For example, PEDAL Portobello Transition Town's focus on land reform partly derives from the same basic considerations of equity that have motivated UK-based campaigns on land access for several centuries.[306] It is further inspired by the broader Scottish movement for community land buyouts, which have positive benefits for community resilience,[307] and those

303 Berkes, F. & C. Folke. 2002. Back to the future: ecosystem dynamics and local knowledge. Pp. 121-146 in Gunderson, L. & C.S. Holling (eds.), *Panarchy*. Washington DC: Island Press.

304 Berkes, F., J. Colding & C. Folke (eds.), 2003. *Navigating Social and Ecological Systems. Building resilience for complexity and change.* Cambridge University Press.

305 Hopkins, R., 2010. *Localisation and Resilience at the Local level: the Case of Transition Town Totnes (Devon, UK)*. Ph.D. thesis, Plymouth University.

306 Shoard, M., 1997. *This Land is Our Land: the Struggle for Britain's Countryside*. London: Gaia Books.

307 Skerrat, S., 2011. *Community Land Ownership and Community Resilience*. Edinburgh: Scottish Agricultural College.

of indigenous groups and other users seeking self-determination through legally sanctioned rights to operate common property regimes. The protection of existing land-based commons, and creation of new ones, and the consequent resistance to increasing consolidation of land ownership and hence access to productive resources, is perhaps the most fundamental of all the outward-facing tasks that Transition and other social justice movements are currently undertaking.

Another way in which Transition practice is moving beyond localism is that around community ownership and management of energy-generating infrastructure. This at one point received significant state support in the UK,[308] where it was initially viewed largely as a means to promote public acceptance of renewable energy technologies,[309] and more recently as a remedy for the contraction of state services associated with the 'localism' agenda of the coalition government that came to power in 2010.[310] More important is the extent to which it increases the potential for active dissent against dominant energy security and climate security agendas.[311] An emerging 'energy commons' allows co-users to express, in their choices of generation technologies and allocation of energy, revenues and other benefits, their own values rather than those of powerful corporate actors.[312] Its existence has demonstrated positive consequences in terms of both empowerment[313] and resilience.[314] Much of the new cultural commons of documentation, experience, knowledge and expertise that supports this arose outside the Transition movement itself: in community renewable energy movements in Denmark,[315] Sweden[316] and Austria,[317] and further back the energy co-operatives responsible for

308 https://www.gov.uk/government/publications/community-energy-strategy. Accessed May 3rd 2014.

309 Walker, G., Hunter, S., Devine-Wright, P. and Evans, B., Fay, H., 2007. Harnessing Community Energies: Explaining and Evaluating Community-Based Localism in Renewable Energy in the UK. Global Environmental Politics 7(2): 64-82.

310 Dawson, W., McCallum, N., Chapple, A., Unwin, E., Lloyd, S. and Fletcher, L., 2011. Funding Revolution: A guide to establishing and running low carbon community revolving funds. London: Forum for the Future and Bates, Wells & Braithwaite Solicitors.

311 Abramsky, K., 2007. Accelerated and far-reaching transition to renewable energies. Why, what, how and by whom? Building New Alliances. World Council for Renewable Energy. Available at: http://www.wcre.de/en/images/stories/pdf/Abramsky_Accelerated_Transition_apr07.pdf. Walker, G. & N. Cass, 2008. Carbon reduction, 'the public' and renewable energy: engaging with socio-technical configurations. Area 39(4): 458-469. Butler, C., S. Darby, T. Henfrey, R. Hoggett & N. Hole, 2012. People and Communities in Energy Security in Butler, C. & J. Watson (eds.) New Challenges in Energy Security – the UK in a Multipolar World. London: Palgrave MacMillan.

312 Wolsink, M., 2011. The research agenda on social acceptance of distributed generation in smartgrids: Renewable as common pool resources. Renewable and Sustainable Energy Reviews 16(1): 822-836.

313 Hathaway, K., 2010. Community Power Empowers. Making Community, Co-operative and Municipal Renewable Energy Happen – lessons from across Europe. London: Urban Forum.

314 Gubbins, N., 2010. The Role of Community Energy Schemes in Supporting Community Resilience. York: Joseph Rowntree Foundation. Harnmeijer, A., J. Harnmeijer, N. McEwen & V. Bhopal, 2012. 'A report on community renewable energy in Scotland', Sustainable Community Energy Network, http://bit.ly/2gTJSzP.

315 Van Est, R., 1999. Winds of Change. A comparative study of the politics of wind energy innovation in California and Denmark. Utrecht: International Books.

316 Henning A., 2000. Ambiguous artefacts. Solar collectors in Swedish contexts. On processes of cultural modification. Stockholm University: Stockholm Studies in Social Anthropology 44.

317 Ornetzeder, M. & H. Rohracher, 2006. User-led Innovations and Participation Processes: Lessons from sustainable energy technologies. Energy Policy 34(2): 138-150.

Figure 4.2.3 – *Community Forest, Lebensgarten, Steyerberg. Credit: Gesa Maschkowski.*

electrification of rural areas across much of the US. Across Europe, key pioneers adapted this knowledge to their national and local contexts.[318] For example, in England and Wales projects such as Baywind Energy Cooperative, Awel Amen Tawe, and West Oxford Community Energy created a common pool of nationally relevant knowledge on which similar projects, many originating in Transition initiatives, have drawn upon this and in turn enriched it as they break new ground and share their learning.[319] The effectiveness of this cultural commons in supporting the establishment and growth of new physical commons (in the form of energy-generating capacity) depends on critical interactions across scales.[320] Particularly important are the existence of suitable financing mechanisms, legal conditions and revenue structures.[321]

318 Hathaway, K., 2010. *Community Power Empowers. Making Community, Co-operative and Municipal Renewable Energy Happen – lessons from across Europe.* London: Urban Forum.

319 For example see *http://www.bristolenergynetwork.org/strategy.* Accessed May 3rd 2014.

320 Avelino, F., Bosman, R. , Frantzeskaki, N., Akerboom, S., Boontje, P., Hoffman, J., Paradies, G., Pel, B. Scholten, D., and Wittmayer, J., 2014. *The (Self-)Governance of Community Energy: Challenges & Prospects.* DRIFT PRACTICE BRIEF nr. E 2014.01, Rotterdam: DRIFT.

321 Hoggett, R., 2010. *The Opportunities and Barriers for Communities to secure at-risk Finance for the Development of Revenue-generating Renewable Energy Projects.* M.Sc. dissertation, Exeter University.

Reclaiming public control over the production of money is another significant step taken by transition communities that heralds broader change given its challenge to the banks' monopoly of money that caused the financial crisis.[322] Alternative financial systems in which money operates as a public good are able to reflect very different values, and have very different consequences,[323] and could be systematically linked to the creation, preservation and nurturing of both community and cultural commons. Several Transition groups have issued complementary currencies, drawing upon, and enhancing, broader pre-existing knowledge commons.[324] In 2010 Transition Bristol (supported by Transition Network, the New Economics Foundation, the Tudor Trust, and others with experience of complementary currencies, including within Transition groups) helped establish a broad alliance of people in Bristol who wanted to set up a complementary currency at the scale of this major city and its surrounding bioregion. They formed a Community Interest Company and in 2012 the Bristol Pound was launched in partnership with Bristol Credit Union.[325] The scheme, which is owned by its members, explicitly seeks to democratise the creation and use of money.[326] Ambitious economic projects of this type force engagement with the public and institutions far beyond the movement itself. A related example is Occupy Wall Street's 'Rolling Jubilee' campaign that buys people's medical debts for five percent of their value and then cancels them.

The commons-building project associated with Transition is still in its infancy, and far from complete. It also has many gaps. For example, very few Transition groups are actively working on water issues (the Netherlands national Transition hub is one notable exception), despite its basic importance as a resource and potential vulnerability to climate change, and the global prominence of struggles over privatisation of water supplies.[327] Elsewhere, the Great Lakes Commons Project[328] is a superb model for linking multiple perspectives on water issues through a commons approach. It does this by promoting a sense of collective responsibility for the ecological health of the Great Lakes among resident communities, and the need to combine protests against profit-seeking by those in power with the assertion of commons-based governance in the face of new threats from fracking, radioactive waste shipments, copper sulphide mining and invasive species.[329]

322 Mellor, M., 2010. *The Future of Money*. London: Pluto Press.

323 Douthwaite, R., 2000. *The Ecology of Money*. Totnes: Green Books. Stokes, P.E., 2009. *Money and Soul: a New Balance Between Finance and Feelings*. Totnes: Green Books. Lietaer, B., C. Arnsperger, S. Goerner & S. Brunnhuber, 2012. *Money and Sustainability: the missing link*. Axminster: Triarchy Press.

324 North, P., 2010. *Local Money: how to make it happen in your community*. Totnes: Green Books.

325 *http://bristolpound.org/*. Accessed November 27th 2013.

326 *http://www.reconomy.org/wp-content/uploads/2013/07/Bristol_Pound.pdf*. Accessed November 27th 2013.

327 Brennan, B., O. Hoedeman, P. Terhorst, S. Kishimoto, B. Balanyá, 2005. *Reclaiming Public Water: Achievements, struggles and visions from around the world*. Amsterdam: Transnational Institute and Corporate Europe Observatory.

328 *http://www.greatlakescommons.org/*. Accessed November 27th 2013.

329 *http://bollier.org/blog/beyond-zombie-environmentalism-great-lakes-commons*. Accessed May 3rd 2014.

4.2.6. A Politics of Place Versus the Politics of Powerlessness

Transition has been criticised for apparently being apolitical,[330] but this is at best only partly true. The agenda it puts forward is one of far-reaching social, cultural and economic change, based around the ongoing creation and expansion of new commons, and deeply subversive of established political and economic orders. The fact that these goals are not overtly politicised allows Transition to permeate a broad range of activities, groups and social contexts not normally associated with radical politics. Its lack of affiliation with any recognised, or recognisable, political creed, can also allow it to be a Trojan horse:[331] a vehicle for the acceptance in and by mainstream politics of radical ideas, framed in such a way as to appear more compatible with dominant political agendas. Transition's standing in the eyes of the UK's former labour government was unprecedented for a grassroots movement of its type. Following that government's replacement with a Conservative-led coalition in 2010, apparent resonance with that government's 'localism' agenda – and wish to give this agenda credibility and divert attention from its retrogressive aspects - masked the difference between this and practical programmes in localisation that may be deeply and productively subversive of neoliberal orthodoxies.[332]

Not being overtly politicised may not remain an option for Transition as it outgrows its focus on discrete localisation initiatives in particular communities. It currently pays little overt attention to, for example, the climate security agenda, but while Transition groups patiently go about their work, globally powerful actors have already imagined their desired future, and are ruthlessly creating it.[333] The two visions are incompatible: the global economy depends on commons, but commons cannot survive its relentless expansion.

The momentum of Transition in its current direction will depend on its success not only in creating alternatives to the global economic system but also in bolstering them through appropriate forms of political support at national and international scales.[334] The way this is likely to play out will be through Transition becoming ever more engaged in the movement of movements that is seeking to resist economic growth and the capture of resources by the few. However, in the process it will be crucial that Transition doesn't lose its place-based focus, a focus which for Transitioners – as for commoners the world over – is about the assertion of what really matters against the insistence that we are all just a point on the grid of extraction, production, consumption and waste. More fundamentally, Transition maintaining this creative focus helps us to resist the tendency to become defined by that which we oppose.

---

330 Chatterton, P. & A. Cutler, 2008. *The Rocky Road to a Real Transition: the Transition Towns movement and what it means for social change.* Trapese Collective. *http://trapese.clearerchannel.org/resources/rocky-road-a5-web.pdf* Accessed October 15th 2012.

331 Leach, M., I. Scoones & A. Stirling, 2010. *Dynamic Sustainabilities.* London: Earthscan. P. 100.

332 *http://transitionculture.org/2010/07/30/localism-or-localisation-defining-our-terms/* Accessed Sept 27th 2012.

333 See various chapters in Buxton, N. & B. Hayes, 2015. *The Secure and the Dispossessed.* London: Pluto Press.

334 A current development of this type is the involvement of Transition Network and national Transition hubs in many EU countries in the establishment of a new EU-wide network for cooperation called ECOLISE (European Community-Led Initiatives for a Sustainable Europe). *http://www.ecolise.eu*

Figure 4.2.4 – *Permaculture Garden, Steyerberg, Germany. Credit: Gesa Maschkowski.*

A key danger for those opposing enclosure-led agendas lies in the fact that such an agenda flourishes not only on secrecy, but on something that appears quite the opposite: the power of appearing to be all-important.

An instructive example of the contrast between the command and control processes of capitalism, backed up by its security agenda, and that of citizen-led movements is evident in their respective proposals for internationally coordinated action to reduce global carbon emissions. The creation of carbon markets perhaps exemplifies the perverse outcomes of applying the same economic logic responsible for climate change to attempts at mitigation. It effectively amounts to enclosure and privatisation of one of the remaining commons, the capacity of the atmosphere and the rest of the biosphere to absorb and buffer disturbances in the carbon cycle.[335] The rationale is that the power of the market will ensure reduction of greenhouse gas emissions proceeds in the most economically efficient way possible, but the reality is very different. In November 2011, Swiss Bank UBS reported that by 2025 the total cost to consumers of the European Emissions Trading Scheme will reach 210 billion euros, with zero or negligible

335 Carbon Trade Watch, 2003. *The Sky is not the Limit.* Carbon Trade Watch Briefing no. 1. Amsterdam: Transnational Institute. www.carbontradewatch.org/downloads/publications/skytexteng.pdf. Accessed October 15th 2012.

impact on emissions reductions but with large increases in energy costs for households and large increases in profits for the most polluting companies. By contrast, citizens have put forward various alternative proposals for the creation of new, equitable and inclusive management regimes for the atmospheric commons. Most are variants of Cap and Share approaches, ways of implementing Contraction and Convergence models that propose global agreement over a timeline for reduction of emissions to acceptable levels, and mechanisms to achieve this in equitable fashion.[336] Cap and Dividend, for example, would place a clear limit on the amount of carbon (fossil fuels) an economy is allowed to ingest, which is rapidly reduced year on year. Companies importing fossil fuels pay increasingly high levies, whose redistribution to all citizens helps them cope with rising costs of carbon-intensive goods and services. This allows the poorest to benefit the most in the transition, and ensures that non-carbon alternatives become more financially attractive as rapidly as carbon based processes disappear – a contraction of carbon occurs alongside a convergence in wealth.

The 'convergence' part of the model is its equity component.[337] It is applicable not only to climate change, but more generally to granting every person on the planet equal rights in relation to access to resources, services and the benefits of resource exploitation, and equal responsibilities in relation to experience of the negative effects of environmental degradation.[338] Convergence is a useful conceptual tool for bringing considerations of distributional justice into debates on sustainability. However, associated with the need for centralised implementation, it carries a risk of imposing conformity.[339] This raises important questions of how intergovernmental mechanisms can reconcile their role in guiding equitable transitions to sustainability with the need to safeguard and promote diversity. Cultural diversity is a key component of social-ecological resilience and a resource for human survival and adaptability in a dynamic, ecologically diverse world.[340] Beyond its utilitarian value, diversity is an inherently necessary and vital part of human existence, without which our existence would be dramatically impoverished, if indeed it were possible at all.[341] This is perhaps the key issue in the relationship between, on the one hand, global mechanisms for sustainability and climate justice of the type the UN or other intergovernmental organisations might seek to implement, and on the other hand, Transition, Occupy, and other grassroots movements rooted in local solutions and distinctive, place-based forms of economic organisation and cultural expression.

336 http://www.gci.org.uk/. Accessed September 27th 2012.

337 Meyer, A., 2000. Contraction and Convergence. Schumacher Briefing no. 5. Totnes: Green Books.

338 Fortnam, M., Cornell, S. and Parker, J. and the CONVERGE Project Team, 2010. Convergence: how can it be part of the pathway to sustainability? CONVERGE Discussion Paper 1. Department of Earth Sciences, University of Bristol.

339 Pontin J. & Roderick, I., 2007. Converging World: Connecting Communities in Global Change. Schumacher Briefing 13. Totnes: Green Books.

340 Petty, J., 2002. Agri-Culture. Reconnecting People, Land and Nature. London: Earthscan. Berkes, F., J. Colding & C. Folke (eds.), 2003. Navigating Social and Ecological Systems. Building Resilience for Complexity and Change. Cambridge University Press.

341 Harmon, D., 2002. In Light of our Differences: how diversity in nature and culture makes us human. London: Smithsonian Institution Press.

What all of these movements have in common is that they actively oppose, and provide alternatives to, the homogenising tendencies of a global economic system that systematically eliminates ecological and cultural diversity, and then seeks to recreate diversity in the form of different products that are sold back to us as consumers.[342] At a local scale, movements such as Transition actively and vitally draw upon community diversity: there are no disposable people in a Transition initiative; everyone has a part to play. At regional, national and international scales, there are no communities or places anywhere in the world without distinctive, self--determined roles to play in transitions to sustainability. Localisation can only be an effective tool in resilience building if it values and honours local diversity, and promotes cooperation and solidarity through suitable linkages at all levels.[343] Commons regimes make these possibilities tangible. It is imperative for the success of Transition and other movements for democratic transitions to sustainability that legal and political mechanisms to ensure that a core plank of intergovernmental action on climate change is the perpetuation of existing commons and creation of new ones.

4.2.7. Conclusion: Climate, Commons and Global Community

Alliances between Transition and Global South movements opposing top-down models of development[344] would be most effective if supported by appropriate international mechanisms such as those described in previous sections of this chapter. Such international mechanisms are needed to ensure the legal protection of existing commons, promote the creation of new commons, and manage the atmosphere as a global commons in ways that systematically link decarbonisation to increased equity. The combination of implementing appropriate models of contraction and convergence and enforceable mechanisms in international law to protect and extend commons of all kinds would create new synergies between localisation movements in the Global North and empowerment of subaltern communities in the Global South. However the route to such international mechanisms is far more likely to be through individual countries unilaterally implementing policies such as cap and share in order to combat ecological meltdown, including climate change, in ways that reduce inequality and strengthen solidarity.[345]

In line with this approach, a key international intervention that individual Transition initiatives (primarily in the Global North) could make would be to link with and support particular commons regimes in the Global South. This could expand the current focus on social enterprise in the Transition movement to include a focus on building economic, social, cultural and political connections between particular localities in the Global North and South – a globalisation from

342 Hardt, M. & A. Negri, 2000. *Empire.* Cambridge, MA: Harvard University Press. Watson, J. L. (ed.), 1997. *Golden Arches East: McDonalds in East Asia.* Stanford University Press.

343 Lewis, M. & P. Conaty, 2012. *The Resilience Imperative.* Gabriola Island, BC: New Society Publishers. Scott Cato, M., 2013. *The Bioregional Economy.* London: Earthscan.

344 A relevant conversation between Transition originator Rob Hopkins and post-colonial scholar Arturo Escobar is recorded at *http://transitionculture.org/2012/09/28/alternatives-to-development-an-interview-with-arturo-escobar/.* Accessed October 15th 2012.

345 For example, the Scottish initiative Holyrood 350. *http://holyrood350.org/.* Accessed May 3rd 2014.

Figure 4.2.5 – *Totnes Market. Credit: Gesa Maschkowski.*

below that is about restoring localities, but also about confronting the bulldozers that are destroying the interlinking localities that constitute our world.

A distinctive perspective that Transition can bring to the struggle against the enclosure agendas is its understanding of our current dilemma in terms of addiction, which goes beyond the idea that it is about an 'us' and a 'them'. The addict does not just tear down the rest of the world to feed his addiction; his addiction also destroys him. This implies a need to alert those wielding power (which to a greater or lesser extent includes those reading and writing this book) to the effects this has on themselves and their children's futures, as much as to the devastation they are causing to others. Chomsky characteristically pointed out that there is no point telling truth to power as power already knows the truth and is busy concealing it from everyone else. But this can easily become a different form of naïveté if it unnecessarily excludes people by saying they are not worth addressing.

From a Transition perspective the key task is twofold:
1. We need to dis-identify from, and oppose, a system we are all to a greater or lesser extent implicated in and addicted to; and
2. We need to build sustainable communities that can both prefigure and help set the direction we need to all take to diminish the climate and related crises, and cultivate networks of cooperation among these.

This combines proactively *reducing* our addiction to oil and the damaging ecological and social impact of our actions, with *adapting* so that we can stand a better chance of helping each other to weather the storm.

Transition combines the belief that we cannot get through unless we are taking everyone with us with the need to know what we are up against. In some senses we are trying to replace relationships of exploitation with relationships of mutual care and support, and the best way of motivating people to work at creating such community gardens, local energy companies, local money, etc., is to focus on the positive and immediate benefits of working with each other now to create a sustainable future. In another sense, we are seeking to transform a whole system not only through resistance but through leadership, example and insistence that there is a far better way of living: one which has always been available to us.

Part of the reason why the news is full of doom and disasters, iced with news of the wealthy and famous, rather than full of the initiatives and care that is happening right now in the world is because there is nothing more threatening to the powers that be, than demonstrating that other more creative, benign, exciting and ancient ways are not only possible but are happening here and now.

Acknowledgements
This chapter was originally written as part of the project that became the Transnational Institute's book The Secure and the Dispossessed, and is slightly modified from the version that appears on the book website *https://www.tni.org/en/publication/the-secure-and-the-dispossessed*.

The writers thank the friends and colleagues who commented on draft versions, particularly Ben Brangwyn and Katy Fox.

4.3. Enchanting Transition: A Post Colonial Perspective
GLEN DAVID KUECKER

4.3.1. Introduction

As concepts like "transition" increasingly become part of our language and thinking about an uncertain future, we need to ask questions about their meanings and uses. This need is especially important when applied to academics and practitioners from the global minority, as they often use language without unpacking the relations of power nested within the knowledge that constructs each word's particular meanings. Instead, words like transition, when used by those in the global minority, take on an objective quality that makes them appear neutral, commonsensical propositions. The neutrality, however, gives way to a world of subjectivity if we inquire whose transition is at stake when we use the term. When a Transition Town in Europe, for example, thinks transition, that thought may well be distinct from what transition means to an Ecuadorian community fighting against mining or for the Zapatista community of *La Realidad* as it contends with the assassination of one of its school teachers. To explore such distinctions, this chapter provides a set of five case studies, which provide a foundation for thinking about transition and knowledge.

The core proposition advanced maintains that actors in the global minority and global majority hold distinct transition epistemologies. These differences carry importance for thinking about the process of social change in the 21st century, as well as the strategic landscape for contending with the great challenges that lie ahead. In making these arguments, the chapter will first present the case studies, and then enter into theoretical discussion about the different practices and meanings of transition as well as the forms of knowledge embedded within them.

Figure 4.3.1 – Glen Kuecker. Credit: Gesa Maschkowski.

4.3.2. Transition Network: An Overview

The Transition Network came into being in 2006 due to the work of people like Rob Hopkins, Naresh Giangrande, and Peter Lipman. According to its web page, the network's central "role is to inspire, encourage, connect, support and train communities as they self-organise around the Transition model, creating initiatives that rebuild resilience and reduce CO_2 emissions." The focus on the climate crisis also led them to directly engage the challenge of peak oil, especially the need to transition to renewable energy and a low carbon economic system. The network emerged from the process that created Transition Town Totnes, a community in the United Kingdom that began working towards creation of an Energy Descent Action Plan based on research undertaken by two of Rob Hopkins' students at Kinsale Further Education College in 2004 and 2005. The transition concept spread rapidly from these modest origins, a fact that demonstrates the success of the network's attempts at sharing information, creating models for transition, carrying out training events, and otherwise supporting local initiatives. In June 2014, the Transition Network claimed 477 "official" transition communities spanning the world with "hubs" in 16 countries, including projects in 56 different countries. These numbers justify viewing the Transition Network as a significant player in constructing alternative practices for adaptation, transition, and resilience.

The ideas guiding the Transition Network are formally articulated in publications, their web page, and a blog. Published in 2008, Rob Hopkins' *The Transition Handbook: From Oil Dependency to Local Resilience* is perhaps the most important publication.[346] It is organized into three sections, each reflecting the overall philosophy of the network. The first section, entitled "The Head," offers a critique of our hydrocarbon civilization and subsequent need for a systemic de-scaling necessary for lower carbon consumption. The second section, "The Heart," addresses our feeling of being overwhelmed by the grand scale of the changes needed, and affirms the need to maintain an

The transition principles articulate a clear set of values that are informed by practice.

optimistic vision of the future. The third section, "The Hands," presents the network's method for taking action, which puts forward a 12-step transition program. In addition to these concepts, the Transition Network embraces Bill Mollison and David Holmgren's permaculture, and it is heavily influenced by "resilience thinking," as advocated by Walter Reid, David Salt and Brian Walker and the Resilience Alliance.[347] Collaboration with Richard Heinberg [348] and

346 Hopkins, Rob. 2008. *The Transition Handbook : From Oil Dependency to Local Resilience*. Totnes, England: Green.

347 Reid, Walter, David Salt, and Brian Walker. 2006. *Resilience Thinking: Sustaining Ecosystems and People in a Changing World*. Washington, DC: Island Press.

348 Heinberg, Richard. 2010. *Peak Everything: Waking up to the Century of Declines*. Gabriola Island, BC: New Society Publishers.

the Post Carbon Institute,[349] as well as the Schumacher Institute[350] adds a layer of complexity theory to the network's perspective[351].

The transition principles articulate a clear set of values that are informed by practice. They offer a vision matched with an orientation toward the pragmatics of how to do transition. They balance between philosophical statements and instruction manual. The transition principles do claim to be an exportable model, which suggests they need to be critiqued from a power/ knowledge perspective, especially because of the universal claims inherent to many of the propositions. The original transition principles, however, are silent on many issues, especially questions of politics, a point that the network addresses in its exchanges with critics, such as that advanced by the Trapese Collective.[352] This friendly critique maintains that the Transition Network inadequately addresses the root causes of the climate and energy crises, and this limitation amounts to the network's inability to confront powerful forces that bar transition from happening. The network underestimates the scale of change necessary, has a flawed understanding of the type of change we need, and is willing to work with local governments. The Trapese Collective's critique emphasizes that the Transition Network is vulnerable to cooptation, and, if too successful, open repression. Perhaps more important, the Transition Network is silent on its position concerning capitalism, which is the fundamental rule-set to the modern system from which it seeks liberation. Perhaps the Transition Network's greatest asset is its ability to vigorously think through its propositions, and allow them to evolve through active debates with its critiques. A debate about Holmgren's "Crash on Demand," for example, opened theoretical spaces for the Transition Network to consider and clarify its position relative to capitalism, especially what role systemic capitalist crises might play in transition.[353]

> *Perhaps the Transition Network's greatest asset is its ability to vigorously think through its propositions, and allow them to evolve through active debates with its critiques.*

349 http://www.postcarbon.org.

350 http://www.schumacherinstitute.org.uk

351 Heinberg, Richard. 2010. *Peak Everything: Waking up to the Century of Declines*. Gabriola Island, BC: New Society Publishers.

352 Chatterton, Paul and Alice Cutler. 2008. "The Rocky Road to Real Transition: The Transition Towns Movement and What it Means for Social Change." Trapese Collective. For Transition Network's response see Hopkins, Rob. 2008. "'The Rocky Road to a Real Transition': A Review." (May 15). Available at: http://transitionculture.org/2008/05/15/the-rocky-road-to-a-real-transition-by-paul-chatterton-and-alice-cutler-a-review/

353 Holmgren, David. 2013. "Crash on Demand: Welcome to the Brown Tech Future." http://holmgren.com.au/wp-content/uploads/2014/01/Crash-on-demand.pdf

4.3.3. Junín: Fighting for the Forests

Tucked away in the pristine cloud forests of Ecuador's northern Andean mountains, the community of Junín, with fewer than 40 families, has defeated two transnational mining companies that had planned on building a larger-scale, open-pit copper mine within community lands.

The community of Junín, with fewer than 40 families, has defeated two transnational mining companies that had planned on building a larger-scale, open-pit copper mine within community lands.

The people of Junín have endured the violence of paramilitary attacks, animosity from neighbors who support mining, and the hardships of everyday resistance. Their twice-won victory is due to the strength of community, a force more powerful than neoliberal transnational capital.

A mixture of factors makes this community so strong. First, excessive marginalization combined with the experience of being *colonos* (land squatters) to forge a strong sense of autonomy. Second, the agrarian roots of the struggle morphed into an environmental struggle, adding a sense of importance to their resistance. Third, *campesino* (peasant) lifestyle merged with marginalization and the *colono* experience to foster strong familial and communal bonds defined by mutual dependence and reciprocity. Taken together these factors generated an immovably resilient force against the mining companies.

Figure 4.3.2 – Community of Junín Fighting Against Transnational Mining Companies. Credit: Mining Watch, Canada.

In response to the severe economic crises of the 1990s, the Ecuadorian government turned to neoliberal policies advocated by the International Monetary Fund and World Bank. These policies drastically reduced funds for state programs, and significantly aggravated rural poverty. The policies also excessively marginalized communities like Junín, which were left to compete in a global agricultural market place. Economic marginalization, absence of the state and its services joined to make Junín nearly invisible within neoliberal plans. In the void left by the state, people in Junín had to provide for themselves. They were equipped for this task, because that's what they had always done. Most community members are second--generation rural squatters, people who came from other parts of Ecuador in search of land and with it security. They carved their community out of the forests. The *colono* experience was defined by extensive hardships, and generated a fierce mentality of independence. Marginalized and born within the rough and tumble world of *colonos*, people from Junín possessed high levels of autonomy, they controlled the process of making decisions about what mattered most to their lives.

Community autonomy also came with being *campesinos*. The people of Junín work their own land, and they are their own bosses. Their production makes them food secure, but they have limited local and regional markets for selling surplus production. They maintain a balance between food security and relative poverty as defined by the capitalist market system. Being poor by the standards of the capitalist economy means protecting food security by controlling their own land and its method of production is a top objective for people living in Junín. Simply stated, community members do not tolerate threats to their economic autonomy. These agrarian roots to community, however, merged with an environmental frame introduced to them by

> *Reciprocity is the key. Each community member is called up to provide community service when needed, through a rotational labor system called the minga.*

outside actors, namely the Church and NGOs. Junín learned that everyone framed the mining struggle as an environmental issue, and they understood the importance of adapting and adopting this movement frame to their agrarian struggle. The environmental discourse added the important element of defending the forests against destruction to their basic struggle, transforming people in Junín into internationally recognized environmentalists.

Another core to community in Junín is how marginalization and the *colono* experience produced strong bonds of mutual dependence and reciprocity within families and neighbors. With less than 40 families, everyone in Junín knows everybody. They pass through the cycles of life together, and know everyone's human strengths and weaknesses. Many families are tightly interconnected through family ties, which form strong bonds of obligations and duties between community members. Surviving the process of settling Junín within a remote part of Ecuador required high levels of mutual support and dependence. Reciprocity is the key. Each community member is called up to provide community service when needed, through a rotational labor system called the *minga*. It ensures that each community member receives the support needed for surviving in an economy of scarcity. Mutual bonds have a cultural articulation in Junín. It is expressed as an honor code, one that values the "word" of each person. To keep one's word and defend one's honor means fulfilling the requirements of community duties and obligations. In a social setting of relative poverty, a person's most valuable possession is often one's word. This social-cultural system is vital in keeping the commons together.

4.3.4. Café R.E.D.: Three Transitions

A few blocks from the central plaza in Quetzaltenango, Guatemala's second largest city, one can find Café R.E.D., a restaurant created by a group of ex-combatants and migrants returned from their pursuit of the *"sueño Americano"* in *"El Norte"*, also known as the United States. The restaurant also houses a grassroots organization, known as DESGUA (Economic Development for a Sustainable Guatemala). The creators of Café R.E.D. designed it to be a "sustainable development" demonstration project. They organized it around the concept of "solidarity food." The restaurant acts in solidarity with local *campesinos* by purchasing their produce, while DESGUA works with them in capacity building and designing local economies around organic food and permaculture practices. An organic chicken farm in Cajola, about an hour from Quetzaltenango, was one of the first indigenous communities to work with DESGUA, and has several community members working at Café R.E.D. as cooks and waiters. The organizers of Café R.E.D. are inspired by the need to replace the *"sueño Americano"* with the *"sueño*

Figure 4.3.3 – *Café R.E.D.'s website.*

Guatemalteco." Many of these activists experienced the brutalities and indignities of migrating to the United States, and many found themselves lost in the absurdities of consumer culture. Most, but not all of the folks at Café R.E.D. are indigenous people, Mayan K'iche' and Mam speakers, and they sought to overcome the alienation of labor and culture by leaving the United States and returning to their homes and culture. Their project, therefore, carries a significant element of cultural recovery and reconstitution, a challenging prospect in Guatemala, where the scars of historical memory from the ugly violence of the civil war's counterinsurgency policies remain deep.

Most participants in Café R.E.D. and DESGUA grew-up with Guatemala's counterinsurgency war raging throughout the province's indigenous communities. Some of the worst repression was only a couple of hours away. The Guerrilla movement recruited several of the organizers when they were high school age and almost all of them became guerrilla soldiers, something that requires a radical transformation in one's life. One of the activists was one of the original founders of the revolutionary movement and was a guerrilla commander for over 30 years. When the civil war ended in a stalemate that led to the 1996 peace accords, many former guerrilla soldiers became deeply disillusioned with the cause to which they had been dedicated, as well as any hope for constructing a more just Guatemala. Disenchanted, many joined the great migration of the 2000s, which found tens of thousands of Guatemalans crossing the border into Mexico, riding the trains north to where they made

> *The organizers of Café R.E.D. are inspired by the need to replace the "sueño Americano" with the "sueño Guatemalteco." Their project carries a significant element of cultural recovery and reconstitution, a challenging prospect in Guatemala*

the crossing into the United States. Once there, many worked in kitchens, which gave them the skills needed to undertake Café R.E.D. upon their return. They settled in places like Madison, Wisconsin, and Morristown, New Jersey. The migrants fell deep into the rabbit hole of American consumerism, and became very distant from Guatemalan culture and identity, especially those who came from indigenous communities. Many married, had children, and found regular work that generated their first steady income. The migrant experience and subsequent Americanization constituted a second radical transformation. These were significant re-workings of one's contexts, identities and life engagements, seldom experienced by people in the global minority, except maybe those we ask to become soldiers and fight our wars. Overtime, this group of migrants came to know that something was not right as they felt the disorder of the world they were in. Many felt unstable, psychologically. Seeing it as a nightmare, they decided to give up on the *sueño Americano* and returned to Guatemala. The return was challenging, as they once again faced the need to rebuild their lives, and once again cross complex borders of identity, meaning, culture, and everyday life. One of them, Willy Barreno, became dedicated to building "*el sueno Guatemalteco*," which led to the creation of DESGUA and Café R.E.D..

While away, the migrants had kept in contact with their comrades from the armed struggle in Guatemala. They are a tight group, what we might view as a Guatemalan version of the "Band of Brothers." They had skills acquired in the United States, and several of them had worked in restaurants. As DESGUA looked for projects, the idea of creating a restaurant where they could train others their trade became one of their first projects. Thus was born Café R.E.D., and the idea for solidarity food. They dedicated themselves, as they had to the guerrilla movement and the quest for the American Dream, to the project.

Unpacking the Café R.E.D. story tells us something more than an example of sustainable development. It provides us a narrative of transition, one that involves giving up on a known world three times as they became guerrilla soldiers, migrants, and returnees. It also provides a remarkable story of human resilience, and social resilience, one of the great stories of the Guatemalan people. It also provides both a literal and metaphorical story of epistemic translation, as they are not simply border crossers but epistemic crossers. In their multiple and layered struggles for liberation, epistemic fluency makes them profound humans, able to conceptualize alternative worlds and develop the plan for their own transition to them.

4.3.5. Pariet Project: Knowledge of the Ancestors

Originating in the 1990s, the Pariet Project of Papua New Guinea seeks to overcome tribal conflicts over issues like land as a way to better organize in resistance to clear-cutting of forests, commercial fishing, and extractive mining. The core of the effort is an extensive process of cultural recovery. In particular, the project focused on re-constituting historical memory through their oral traditions, a particular knowledge held by tribal elders that connects community to time and space. Tribal elders can reach back 36 generations through their memory and story telling, a living archive of wisdom about how humans relate with one another, nature, and the cosmos. Zureki Maigao, Chairperson of the Pariet Project[354], explains the Pariet Project by stating:

> *This is now our story. I have been Chairperson for fifteen years. We are Pariet. We have been part of the Pariet story for more than fifteen years. It is slow, it is hard, and it is a struggle to find our roots, our ancestors after so much destruction. But we are stronger than ever before. Under [sic] now the tribal lineage of the Ammam people, we stretch far across the mountains and down through to the other side of the mountains. Pariet has helped us to see who we are. We want our independence and our own way to decide what is for our self-betterment. In my role, I speak with authority. Pariet will see us through. I have the Amman blood, and in my blood, I know this is the way.*

354 Nadarajah, Yaso. 2009. Wisini Group of Villages, Morobe Province. *Local-Global* 5: 117.

Nadarajah explains the significance of the Chairman's words:[355]

> *It is both a contingent and choreographed story of contemporary intersections as people look in two directions at once: back to a carefully reconstructed past and forward in modern time to the future that brings together different ways of being. Their story includes a symbolic journey by outsiders to their place, a modern chronicle of a political movement called Pariet, a tribal dance and hidden stories that remain unrevealed here; and it ends with the words of their Chairperson, 'I speak with authority', old and new.*

The geography of 'their own spaces,' is found here in the in-between places of culture and community, the nooks and crannies of life where modernity has not visited, and when it has, it has not won.

An example of the hybridity of Pariet Project's in-between spaces comes from Naup Waup, the founder of the Pariet Project. Waup is from the tribal community of Wisini, a full day's ride, if the "road" is good, from the port city of Lae in Morobe Province, an hour flight from the capital of Port Moresby. When they travel down the mountains to Lae, most villagers take the forest trails, instead of the modern road. Tribal elders selected Waup, upon his birth, to be the next tribal leader. To prepare for the task the community gave him special treatment as he grew up. Knowing that the modern world was finding its way to Wisini, the elders understood that the next leader needed a modern education, so they sent Waup to Lae for formal schooling. He excelled in his classes, and developed what became a charismatic personality, one that captured attention through energy, intensity, and intelligence. Waup gained a scholarship for university study in Australia, at RMIT University in Melbourne. He studied art. Waup returned to Papua New Guinea, living with one foot in the modern city of Lae and the other in Wisini. He had become an expert navigator between wildly divergent cultures, able to translate meanings, contexts, and espistemologies. Tribal elders passed along their knowledge of the ancestors to Waup, a different form of education that prepared him for some day becoming a tribal elder. Taking from the modern world, and knowing the need to preserve tribal ways, Waup launched the Pariet Project.

Waup is an amazing artist. He does prints that articulate dream visions of the other knowledges of his people. They are intense representations of tribal origin myths that involve the original migration of the founding lineage, as well as the truths of the elders. The art is his way of communicating the hybrid world that makes cultural survival so important and so challenging.

4.3.6. Zapatistas
From the mountains of southern Mexico a revolutionary force awoke the world to the possibility of a radical alternative to the triumph of post-cold war neoliberal globalization. Known as the Zapatistas, indigenous peoples of Mayan origin launched an armed rebellion against the "bad government" of the Mexican state and the economic forces of neoliberalism

355 Nadarajah, Yaso. 2009. Wisini Group of Villages, Morobe Province. *Local-Global* 5: 117.

on January 1, 1994. Their initial rebellion, faced with overwhelming military repression from the Mexican military and civil society's demand for peace, transformed into a new form of revolution, one that did not seek the seizure of state power as the tool for revolutionary change, but instead sought to "have a revolution without taking power." The Zapatista project became a well articulated vision for change, one that converged a program for good government and an self-sufficient and equitable economy around the concept of autonomy, which invested the fundamental right to decide within the particular cultural practices of community. The Zapatista revolution was explicitly against neoliberalism, as it saw policies like the North American Free Trade Agreement as the "death sentence" of indigenous people. Their revolution, however, was also anti-capitalist, a point made abundantly clear by their Sixth Declaration of the Lacandon Jungle, which was put forward to the international community in 2005. Their revolution is marked by the longevity of its open rebellion, twenty years and counting, which illustrates the open nature of the rebellion, one that emphasizes process, the "the revolution is the path you walk," over dogma that allows it to be continuously responsive to changing contexts and proactive in formulating new directions in revolutionary struggle. The Zapatistas also stand out because they retain a revolutionary army, one mobilized for the defense of liberated communities throughout Chiapas.

> *The Zapatista project became a well articulated vision for change, one that converged a program for good government and an self-sufficient and equitable economy around the concept of autonomy, which invested the fundamental right to decide within the particular cultural practices of community.*

The Zapatista movement anchors its approaches within the "norms and customs" of indigenous communities. These are shaped by half a millennium of encounter with the modern world, which has had an obsessive vision of indigenous people as barriers to the world of Enlightenment. Their resistance is cultural; it merges the continuity of indigenous ways of being, seeing, thinking, and acting, with a knowledge that comes from selective interaction with modernity, where their norms and customs remain subaltern to the way power is constituted by modernity. The practice of autonomy is a cultural proposition, one that actualizes indigenous cosmovision with the arts of resistance and constituting an alternative world. Rooted in distinct notions of time and space, the cosmovision places emphasis on the human struggle to balance between the universe's forces of order and disorder. Modernity, in this view, wrought centuries of disequilibrium, a profound disordering of the universe, so that their revolutionary struggle is the attempt to bring order back to the universe. These norms and customs make the Zapatista movement highly resilient and mark it as a transcendental proposition.

Figure 4.3.4 – Zapatista Cosmovision: "We Want a World Where All Worlds Belong". Credit: Michael Cucher.

4.3.7. Enchanting Transition

Transition is a core concept for understanding the narrative within each of this chapter's case studies. For the Transition Network, moving to a post hydrocarbon world of re-localized economies means breaking from the logic of modernity. For Junín, the evolution from colonizers to environmentalists defined their transition, one that placed resistance to mining as the way they made meaning of the transition. Three deep transitions – from peasant to guerrilla, guerrilla to *sueño Americano*, and *sueño Americano* to the return migrant's *sueño Guatemalleco* – rest at the core of the solidarity food idea that gave birth to *Café R.E.D.*. In Papua New Guinea, cultural recovery and community organizing constituted the transition embraced by the Pariet Project. The creation of autonomy and its radical re-thinking of revolutionary change defined the great transition of the Zapatistas. From these case studies, we can see that there is no single, homogenous meaning to transition. Instead, they show that transition has multiple meanings that are highly contingent upon historical contexts and human experiences. This diversity points to the importance of asking "whose" transition when we deploy the concept in our theoretical and practical work.

A useful approach for thinking about the diversity of transitions present in this essay's case studies is the Frankfurt School of Sociology's "enchantment" proposition. Adorno and

Horkheimer (2002), in their *Dialectic of Enlightenment*, trace the origins of modernity's dark side as resting with the culture-nature dichotomy created by Bacon and Descartes' "war against nature." Adorno and Horkheimer argue that modern ways of being, seeing, thinking, and act-ing seek to exterminate the enchanted world, the ways of being, seeing, acting, and thinking that existed prior to the war on nature. With their secular world of rationality, moderns live in a disen-chanted world. To overcome the dark side of modernity and leverage the positive aspects of enlightenment rationality, moderns have to experience a process

To overcome the dark side of modernity and leverage the positive aspects of enlight-enment rationality, moderns have to experience a process of re-enchantment.

of re-enchantment. Taking from this framework, we can think of an epistemic continuum that extends between the extremes of enchanted and disenchanted worlds, and where re-enchantment occupies a middle position. We can locate the Transition Network in a space between disenchantment and re-enchantment, while the Pariet Project and Zapatistas would fall toward the enchantment end of the continuum. In this interpretation, *Café R.E.D.* and the community of Junín hold a space within re-enchantment due to their significant engagements in the modern world.

Figure 4.3.5 – Movements' locations on a continuum from enchantment to disenchantment.

Nested within the continuum of transition between forms of enchantment are different forms of knowledge. Through its practices of creating communities of transition, the Transition Network generates knowledge. This knowledge is reproduced through an intentional effort at making its experiences known to a wider public. Knowledge gained from experience is formalized through several mediums, including the Transition Network web page, the blogs of its advocates, and from the intellectual work of researchers. Formalization provides the Transition Network's knowledge with legitimacy that enhances its status. Through repetition and expansion, its knowledge production is embedded within relations of power, because it defines questions that need to be studied, generates the analysis of those problems, and from the analysis proposes actions to be taken. It defines the "facts" to be known about transition, what is "legible" when transition is discussed, and participates in a wider system of understandings shared by similar organizations, but also the wider world of power/knowledge actors, such as UN agencies, the World Bank, and various levels of government. Resilience, for example, is a concept and proposition known to all of these actors, to an extent that there is a level of "common sense," meaning hegemony, within its use. As facts are defined and realities made legible, the Transition Network participates in an erasure of alternatives, it necessarily excludes some possibilities from being "facts" and renders some realities "illegible" as it does its work. These are subtle acts, often committed without awareness or critical perspective, or without malicious intent. They do, however, constitute forms of knowledge and power.

The Transition Network produces a particular type of knowledge about transition. First, they define the problem from the perspective of climate change and peak oil, whereby transition becomes a question of breaking free from hydrocarbon civilization. The "facts" of transition become constituted by the scientific discourse about climate change and energy, as well as the political and economic system sustaining the problem. This construction of transition makes the modern system legible, a process that defines the sequencing of transition itself, a movement from the modern system to an "alternative" system, a new reality that is not yet named except for its reference to the system it is transitioning from, such as the "post-hydrocarbon world."

When we turn to the global majority case studies, however, transition takes on very different meanings. While transition for the global minority means a departure from being modern, for those in the global majority, they are already an alternative. Being an alternative, the global majority case studies act in resistance to modernity either rejecting it entirely or in an effort to engage modernity on its own terms, often in a subaltern fashion that is illegible to the ways

While transition for the global minority means a departure from being modern, for those in the global majority, they are already an alternative. Being an alternative, the global majority case studies act in resistance to modernity either rejecting it entirely or in an effort to engage modernity on its own terms.

moderns constitute knowledge. When these case studies articulate their concerns, climate change and peak oil are seldom mentioned, even though the participants are often well aware of those crises. Juxtaposed to these case studies, the Transition Network's agenda appears to be constituted by the global minority for the global minority. The Transition Network's construction of transition renders illegible the "other knowledges" generated by the global majority that constitute different paths toward the future.

4.3.8. Other Knowledges and the Transmodern World

Post-colonial studies offers promising ideas for the task of understanding the other knowledges generated by communities in the global majority, especially those communities, such as the case studies in earlier sections in this chapter, that retain a significant extent of an enchanted epistemic. Post-colonial studies is useful because of its pre-occupation with understanding how forms of colonial power shape the experience of the colonized upon their formal liberation from colonialism. This perspective allows us to understand how the struggle for liberation is constitutive of the lived reality for most if not all of the global majority. The challenge of liberation, as explored by thinkers like Homi Bhabha, Franz Fanon and Albert Memmi.[356] represents a context distinct from the type of liberation sought by groups in the global minority, such as the Transition Network. The post-colonial condition is a way of being human that has its own ways of seeing and thinking. A significant part of the post-colonial epistemic is the theoretical and practical problem of escaping what appears to be an everlasting condition of having been colonized. The struggle to transcend the colonial marking constitutes the core meaning of transition for the global majority, a meaning that is radically distinct from that constructed by the Transition Network.

The knowledge produced by the other transitions of the global majority, is derived from three key sources: first, from the reality of a community's being post-colonial. second, from the everyday life struggles of marginalized peoples and the third form of knowledge is legacy knowledge, those pre-colonial enchanted epistemologies and cosmologies that survived modernity, whether in fragments or in entirety.

Within post-colonial studies, vigorous debates exist about the theory and practice of liberation for the global majority. Among them are the contested understandings of what it means to be "subaltern," that exist between scholars from the global minority and the global majority,

356 Bhabha, Homi. 1994. *The Location of Culture*. New York: Routledge. Fanon, Frantz. 1965. *The Wretched of the Earth*. New York: Grove. Originally published 1961; Fanon, Frantz. 1967. *Black Skin, White Masks*. New York: Grove. Originally published in 1952. Memmi, Albert. 1965. *Colonizer and Colonized*. New York: Orion. Originally published in 1957

such as the debate that exists within the Latin American Subaltern Studies group. Their divide was between scholars from the global minority, such as Florencia Mallon,[357] who tended to have an ironically eurocentric critique of eurocentrism, and those from the global majority (especially a group of Latin Americanists, led by Anibal Quijano and Quijano and Ennis, Enrique Dussel, Dussel, Krauel and Tuma and Dussel, Moraña, and Jáuregui, Walter Mignolo, and Rámon Grosfoguel), who seek an understanding of the post-colonial condition that is derived from what they see to be a truly subaltern perspective – its "other knowledges" – created by the post-colonial experience.[358] Echoing the Frankfurt School's thesis that modernity exterminates the enchanted, their work explains how colonial repression aimed to eradicate other knowledges, and replace them with the modern epistemic. Quijano writes, "The repression fell, above all, over the modes of knowing, of producing knowledge, of producing perspectives, images and systems of images, symbols, modes of signification, over the resources, patterns, and instruments of formalized and objectivised expression, intellectual or visual."[359] Transition for the global majority, therefore, requires untangling all of the complex webs of deeply embedded and interrelated relations of power within economy, authority, gender, and subjectivity and knowledge that are each defined by the modernity's inequities and inequalities, but also the struggle to remain enchanted.[360]

The knowledge produced by the other transitions of the global majority is derived from three key sources. The first knowledge derives from the reality of a community's being post-colonial.

357 Mallon, Florencia. 1994. "The Promise and Dilemma of Subaltern Studies: Perspectives from Latin American History." *American Historical Review*. Vol. 99; No. 5. (December): 1491-1515.

358 Quijano, Aníbal. 2007. "Coloniality and Modernity/Rationality." *Cultural Studies*. Vol. 21; Nos. 2-3. (March/May): 168-178; Quijano, Aníbal and Michael Ennis. 2000. "Coloniality of Power, Eurocentrism, and Latin America." *Neplanta: Views from South*. Vol. 1; No. 3: 533-580; Dussel, Enrique. 2004."Deconstruction of the Concept of 'Tolerance': From Intolerance to Solidarity." *Constellations*. Vol. 11; No. 3: 326-333; Dussel, Enrique. 2002. "World System and 'Transmodernity.'" *Neplanta: Views from the South*. Vol. 3; No. 2: 221-244.; Dussel, Enrique. 2001. *Towards an Unknown Marx: A Commentary on the Manuscripts of 1861–63*. London: Routledge; Dussel, Enrique. 1998. Beyond Eurocentrism: The World-System and the Limits of Modernity." In Fredric Jameson and Masao Miyoshi. Editors. *The Cultures of Globalization*. Durham, NC: Duke University Press, 1998.; Dussel, Enrique. 1996. *The Underside of Modernity: Apel, Ricoeur, Rorty, Taylor and the Philosophy of Liberation*. Atlantic Highlands, NJ: Humanities Press International, 1996; Dussel, Enrique. 1995. *The Invention of the Americas: The Eclipse of the "Other" and the Myth of Modernity*. New York: Continuum, 1995; Dussel, Enrique. 1980. *Philosophy of Liberation*. Maryknoll, NY: Orbis Books, 1980; Dussel, Enrique, Javier Krauel and Virginia Tuma. 2000. "Europe, Modernity, and Eurocentrism." *Neplanta: Views from the South*. Vol. 1; No. 3: 465-478; Dussel, Enrique, Mabel Moraña, and Carlos Jáuregui. 2008. Editors. *Coloniality at Large: Latin America and the Postcolonial Debate*. Durham, NC: Duke University Press.; Mignolo, Walter. 2005. *The Idea of Latin America*. Oxford: Blackwell, Mignolo, Walter. 2002. "The Geopolitics of Knowledge and the Colonial Difference." *The South Atlatnic Quarterly*. Vol. 101; No. 1: 57-96, Mignolo, Walter. 2000a. *Local Histories/Global Designs: Coloniality, Subaltern Knowledge, and Border Thinking*. New Jersey: Princeton University Press, Mignolo, Walter. 2000b. "The Many Faces of Cosmo-polis: Border Thinking and Critical Cosmopolitanism." *Public Culture*. Vol. 12; No. 3: 721-748, Mignolo, Walter. 1995. *The Darker Side of the Renaissance: Literacy, Territoriality, and Colonization*. Ann Arbor: University of Michigan Press.; Grosfoguel, Rámon. 2008. "Transmodernity, Border Thinking, and Global Coloniality: Decolonizing Political Economy and Postcolonial Studies." *Eurozine*. (July): 1-23. *http://www.eurozine.com/pdf/2008-07-04-grosfoguel-en.pdf* (accessed July 2010).

359 Quijano, Aníbal. 2007. "Coloniality and Modernity/Rationality". *Cultural Studies*. Vol. 21; Nos. 2-3. (March/May): 169.

360 Mignolo, W. D., 2007. "Introduction: Coloniality of power and de-colonial thinking." *Cultural Studies* Vol. 21; Nos. 2-3. (March/May): 157.

The second knowledge is produced from the everyday life struggles of marginalized peoples. It is the knowledge about being human acquired from the internal conflicts and dilemmas of people in resistance, the struggle to retain an enchanted world. The third form of knowledge is legacy knowledge, those pre-colonial enchanted epistemologies and cosmologies that survived modernity, whether in fragments or in entirety. These other knowledges are characterized by their diversity. They offer a "pluriversal" truth as against the modernity's universal truth. The post-colonial embrace of difference is rooted in its critique of modern rationality that is informed by the Frankfurt School's critical theory.[361] It finds the universalism of the modernity to be an oppressive mechanism that obliterates difference through colonial relations of power.[362] The post-colonial instead sees diversity as anchored in the way many communities in the global majority have de-centered, localized, and plural ways of being, acting, and thinking that have persisted despite the homogenizing ways modernity has historically constituted difference as binary oppositions such as modern versus traditional or developed versus backward.

The post-colonial condition constitutes a different context and meaning about 21st century transition when juxtaposed to the knowledge generated by the Transition Network. Some post-colonial theorists call this distinct knowledge "transmodernity," and see it as a means for global majority post-colonials to transcend modernity, as against transition from it. Dussel, for example, states that transmodernity "will have a creative function of great significance in the twenty-first century." The diversity of transmoderns, their lived experiences, and legacy knowledges that form the base of the transmodern epistemic is uniquely matched for the creativity and experimentation needed for weathering the perfect storm of 21st century challenges. It is the time of the global majority, when the meek will inherit the earth. Dussel states:[363]

> "modernity's recent impact on the planet's multiple cultures (Chinese, Southeast Asian, Hindu, Islamic, Bantu, Latin American) produced a varied 'reply' by all of them to the modern 'challenge.' Renewed, they are now erupting on a cultural horizon 'beyond' modernity. I call the reality of that fertile multicultural moment 'trans'-'modernity'."

Beyond modernity the transmoderns return "to their status as actors in the history of the world-system."[364] As reconstituted agents unmarked by the condition of colonialism, transmoderns will be their own protagonists in the making of 21st century history. Transmoderns "retain an immense capacity for and reserve of cultural invention essential for humanity's

361 Dallmayr, Fred. 2004. "The Underside of Modernity: Adorno, Heidegger, and Dussel." *Constellations*. Vol. 11; No. 1: 102-120. Mignolo, W. D., 2007. "Introduction: Coloniality of power and de-colonial thinking." *Cultural Studies* Vol. 21; Nos. 2-3. (March/May): 155.

362 Quijano, Aníbal. 2007."Coloniality and Modernity/Rationality." *Cultural Studies*. Vol. 21; Nos. 2-3. (March/May): 168-178. Dussel, Enrique. 2002. "World System and 'Transmodernity.'" *Neplanta: Views from the South*. Vol. 3; No. 2: 221.

363 *Ibid*. p. 221

364 *Ibid*. p 224

survival," Dussel states.[365] His version of transmodernity promises a new humanism, where "these cultures, in their full creative potential.... constitute a more human and complex world, more passionate and diverse, a manifestation of the fecundity that the human species has shown for millennia."[366]

4.3.9. Conclusion: Is a New Humanism Possible?

The Transition Network and the global majority case studies considered in this chapter share several points of common ground. Each constitutes a path toward escaping the death spiral of modernity through the creation of an alternative way of being, seeing, thinking, and acting. All are creative responses to the deepening crises of modernity that carry important but nuanced aspects of the type of resilience essential for weathering the perfect storm of the 21st century. There are also important distinctions that mark the Transition Network alternative as a substantially different proposition when compared to the global majority cases. The Transition Network occupies a position of privilege and relative power that comes with being a global minority movement. While facing challenges of counter hegemony, the Transition Network has yet to face the oppression and repression that alternative paths in the global majority routinely encounter. While each produces knowledge that generates power, the Transition Network significantly operates within the power/knowledge regime of modernity. At least one foot remains firmly embedded in the disenchanted world. It does not experience the post-colonial context of the global majority that produces their other knowledges and the proposition of transmodernity. In the final analysis, the Transition Network exists in a distinct epistemic defined by a problematic of disenchantment, as against one coming from the enchanted world struggling for liberation from colonialism and its legacy.

Among the many unresolved questions explored in this chpater, and one that softly lurks within each section, concerns one of the greatest offerings made by the Enlightenment, the challenge of solidarity among humanity in the struggle for liberation. I have explored the question of solidarity in an essay,[367] "Academic Activism and the Socially Just University." It undertakes a critical reflection on my own efforts at solidarity, and through it the essay postulates that solidarity may not be possible. Despite that pessimism, the intent was more hopeful in that I view solidarity as an on-going struggle toward a goal that will always be beyond our reach. It is the effort at reaching the goal that carries the human potential for liberation, and it is a necessary effort, especially now that we face a new historical epoch defined not by the brutalities of neoliberal globalization, but by the prospects of modernity's collapse. The collapsing structures offer us the opportunity to innovate, create, and transition to something more human. It is the time to realize the imaginary of solidarity, and through that effort to create a new humanism for the 21st century.

365 *Ibid*. p. 235

366 *Ibid*. p. 237

367 Kuecker, Glen David. "Academic Activism and the Socially Just University." Pp. 42-55. In: Kathleen Skubikowski, Catherine Wright, and Roman Graf. Editors. *Social Justice Education: Inviting Faculty to Transform Their Institutions*. Sterling, VA: Stylus Publications, 2010.

Dussel maintains that transmoderns have the capacity to create the new humanism. In that project, what role do transitioners in the global minority play? One possible answer comes from an interview Rob Hopkins had with Arturo Escobar, a post-colonial theorist who is part of the group that theorized transmodernity. Escobar states,[368]

> *"The concept, the practice of Transition that we use for different parts of the world, we have to take into account that they will be inter-cultural conversations, inter-epistemic conversations, different knowledge is going to be involved, and those require translation. Translation across knowledges, across cultures, across histories, across different ways of being negatively affected by globalisation, across levels of privilege and so forth."*

That undertaking requires deployment of critical theory to unpack the hegemonic uses of key concepts, such as "transition" so that we can destabilize the knowledge embedded in their meanings. This destabilization opens spaces for translation between epistemics and traditions, especially between the enchanted, re-enchanted, and disenchanted worlds. Escobar's provocative statement joins other post-colonial theorists[369] in proposing the destabilizing of knowledge in order to decenter it so that we may become "border crossers."

368 Hopkins, Robert. 2012. "Alternatives to Development: An Interview with Arturo Escobar."
     September 28, 2012. Availabe at: *https://www.transitionculture.org/2012/09/28/
     alternatives-to-development-an-interview-with-arturo-escobar/*

369 Mohanty, Chandra. 2003. Feminism Without Borders: Decolonizing Theory, Practicing Solidarity.
     Durham: Duke University Press; Smith, Linda Tuhiwai. 1999. *Decolonizing Methodologies: Research and
     Indigenous Peoples*. New York: Zed Books; Anzaldúa, Gloria. 1999. *Borderlands/la frontera: The New Mestiza*
     (2nd ed.). San Francisco: Aunt Lute Books.

Index

The Editors

Dr. Thomas Henfrey is Senior Researcher at the Schumacher Institute and Research Fellow in the Centre for Ecology, Evolution and Environmental Change (cE3c) at Lisbon University. He previously conducted PhD research on indigenous forest management and lectured in the Anthropology Department at Durham University. Tom co-founded an ecovillage in Southern Spain, holds a permaculture design certificate, and actively collaborates on research and learning with Ecolise, Transition Network and the Permaculture Association (Britain).

Gesa Maschkowski is a nutritionist with extensive professional experience in sustainability and nutrition education. She co-founded the Transition Initiative Bonn-im-Wandel e. V., and the first Community Supported Agriculture project in Bonn. Gesa loves to facilitate change processes, including by working as Transition Trainer. She also has a passion for photography and participatory video, making transition visible in Bonn and elsewhere. Currently, she is a Ph.D. student at the University of Bonn. Her research interests include resilience, salutogenesis and transdisciplinary research. She blogs about Transition research on the website of the German Transition Network and is member of the German team of Transition Trainers.

Dr. Gil Penha-Lopes, an Environmental Scientist by training, is an Invited Professor in the Science Faculty at Lisbon University and Senior Researcher at the Centre for Ecology, Evolution and Environmental Change (cE3c). He coordinates Lisbon University's involvement in the EU-funded FP7 BASE Project on bottom-up approaches to climate change adaptation, and the ClimAdaPT.Local project, which collaborates with 26 municipal councils across Portugal to devise local climate change adaptation strategies. A certified Transition Network Trainer, Gil contributes and promotes Permaculture and Transition (Towns) research in Portugal and Worldwide.

Community-led Transformations
Series Editors: Thomas Henfrey and Gil Penha-Lopes

The Community-led Transformations book series is a collaboration between members of the ECOLISE Network of European community-led sustainability initiatives and independent sustainability publishers, Permanent Publications. It communicates intellectually rigorous thinking from the interface of research and practice, supporting local action and improved formulation and delivery of policy towards a fairer, more sustainable world.

Volume 1: Permaculture and Climate Change Adaptation. T. Henfrey & G. Penha-Lopes

Volume 2: Resilience, Community Action and Societal Transformation. T. Henfrey, G. Maschkowski & G. Penha-Lopes

Books to empower your head, heart and hands

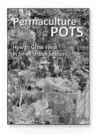 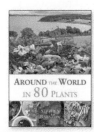   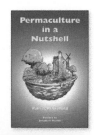

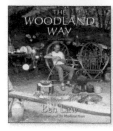 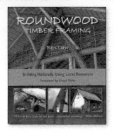  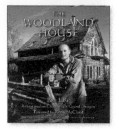

For our full range of titles, to purchase books
and to sign up to our free enewsletter see:

www.permanentpublications.co.uk

also available in North America from:
www.chelseagreen.com/permanentpublications

Subscribe to
permaculture International
practical solutions beyond sustainability

Permaculture Magazine International offers tried and tested ways of creating flexible, low cost approaches to sustainable living

Print subscribers have FREE digital and app access to over 25 years of back issues

To subscribe, check our daily updates and to sign up to our free eNewsletter see:

www.permaculture.co.uk

To subscribe to our North American specific edition see:

www.permaculturemag.org